数学的基础

Math Foundations

中英双语指南 A Bilingual Guide

Table of Contents

前言 Preface

This bilingual book, in English and Chinese, is for people who have already learned math in one language and have the need to understand or teach in the other language. Students from China studying abroad, Chinese parents who need to teach their English speaking children, and people who wish to work in China, will all benefit from this book.

这本中英文双语书适用于已经用一种语言学过数学，但需要用另一种语言理解或教学的人。在国外学习的中国学生、需要教说英语的孩子的中国父母、以及希望在中国工作的人都将从本书中受益。

We are not re-teaching basic math skills. We also assume readers already have basic language skills in both languages. In this book, the goal is to make it easy and efficient for the reader to learn math vocabulary in another language. We make everything brief to save the reader time. Whenever possible we'll present in tables and pictures to make comparisons straight forward. Readers will find it is very easy to use this book to quickly review elementary math, if you have already forgotten the details you learned years ago. It might bring back some of your fun childhood memories as well.

我们不是在重新教授基本数学技能，我们还假定读者已经具备两种语言的基本技能。在本书中，我们的目标是让读者能够轻松高效地学习另一种语言里的数学词汇。我们尽量简明扼要，以节省读者的时间。在可能的情况下，我们会以表格和图片的形式进行展示，以便直接双语对照。如果你已经忘记了多年前所学的细节，用这本书来快速复习初等数学将轻松自如。甚至，它还能唤起你一些有趣的童年回忆。

This book covers mathematics learned in elementary school and middle school, the foundation of further studies. It is the first in a series of Chinese-English bilingual books. The series will expand to a wide range of topics, including more advanced math such as Algebra and Calculus. Interested readers please visit our web site for more information.

这本书涵盖了小学和初中学到的数学知识，是进一步学习的基础。它是一个中英双语书籍系列中的第一本。系列将扩展到广泛的话题，包括代数和微积分等更高深的数学。有兴趣的读者请访问我们的网站以了解更多信息。

1 数字系统 Numeral System

从毛茸茸的小鸡到庞大的大象，很多动物知道数数，但只有人类还发明了数学。经过几万年一代代人的努力，数学发展到了今天让人陶醉的程度。这一切都离不开文字，用符号表示数字。让我们就从各种文字使用的数字开始说起（表1）。

From fluffy chicks to gigantic elephants, many animals know how to count, but only humans have invented mathematics. Through ten thousand years, generation after generation, we have made math enchanting. All of this is inseparable from words, using symbols to represent numbers. Let's start with the numbers used in various languages (Table 1).

Table 1. *Numbers in different languages* **表1.** 各种语言的数字

印度-阿拉伯 Hindu–Arabic	婆罗米 Brahmi	罗马 Roman	中文 Chinese	大写中文 Capitalized	英语 English
0	·		〇	零	Zero
1	–	I	一	壹	One
2	=	II	二	贰	Two
3	≡	III	三	叁	Three
4	Ɏ	IV	四	肆	Four
5	h	V	五	伍	Five
6	ε	VI	六	陆	Six
7	?	VII	七	柒	Seven
8	?	VIII	八	捌	Eight
9	?	IX	九	玖	Nine

今天最广泛使用的数字是阿拉伯数字。阿拉伯数字最早是在印度开始使用的，后来经阿拉伯传遍全球，如今成为我们共同的数学语言。不管你的母语是什么，大家都能看懂阿拉伯数字。

Today, the most widely used numerical symbols are Arabic numbers. Arabic numbers were first developed in India, spread all around the

6

world by Arabs, and have become the common math language of all of us. Whatever your native language is, we can all understand Arabic numbers.

各文明还是保留了他们自己语言中的数字，例如英语的 zero, one, two, three 等。中文也使用零、一、二、三等。中文中还会见到笔画很多的大写数字，如壹、贰、叁、肆、伍等。这些数因为不易被篡改，银行仍然使用，比如人民币上你会发现这些大写的数字，5 写成伍（图 1）。类似的在英语国家，纸币和银行支票也必须同时使用英语数字，比如 5 美元纸币上一定有 FIVE 这个字（图 1）。

Civilizations still retain numbers in their own languages, such as zero, one, two, three, etc. in English. Chinese also uses 零、一、二、三 and so on. In Chinese, you will also see capitalized numbers with a lot of strokes, such as 壹、贰、叁、肆、伍 etc. These numbers are still used by banks because they are not easily tampered with. For example, you will find these capitalized numbers on the Renminbi, where 5 is written as 伍 (Figure 1). Similarly in English-speaking countries, banknotes and bank checks must also use English numerals, for example, a $5 bill also has the word FIVE on it (Figure 1).

你能找到伍吗?

Can you spy FIVE?

Figure 1. Capitalized numbers on currency **图 1.** 钱上的大写数字

2 十进制 Decimal

位值 Place Value

有了 0 到 9 之后我们可以用十进制的系统来表示大于 9 的数字，其原理是基于一个叫做位值的概念——一个数字的大小，不仅取决于数字本身，还取决于数字所处的位置。有了这个系统，我们就可以重复利用 0-9 这十个字符，用以表示任何数字。

With 0 to 9 we can use the decimal system to represent numbers greater than 9, based on the concept called place value, that is the value of a number is determined not only by the number itself, but also by the position of the number. With this system, we can reuse the ten characters 0-9 to represent any number.

在我们讲位值的概念前我们先熟悉一下要用到的中英词汇。首先是十进制的单位，中文用个十百千万——个常常省略不提（表 2）。这些单位可以循环重复使用，如十万、百万等等。万万是一个新单位亿，万亿是兆。英语中用 One Thousand，Ten Thousand，Hundred Thousand, etc。Thousand Thousand 是 Million，Thousand Million 是 Billion，Thousand Billion 是 Trillion。因此英语中在书写阿拉伯数字时会将整数以千为单位，分成三个一组，用 "," 作为千位分隔符。例如 331,449,281。千位分隔符在中文中也被广泛采用，而不是以万为单位引入万位分隔符。K 常常会被用作 Thousands 的简写，5000 会被写成 5K。

Before we talk about the concept of place value let's familiarize ourselves with the Chinese and English vocabulary to be used. First of all, the units of decimals, Chinese use 个十百千万, 个 is often omitted (Table 2). These units can be combined and used repeatedly, for example 十万, 百万 and so on. 万万 is a new unit 亿, 万亿 is 兆. English uses One Thousand, Ten Thousand, Hundred Thousand, etc. Thousand Thousand is Million, Thousand Million is Billion, Thousand Billion is Trillion. Therefore, in English, when writing Arabic numerals, the whole number will be divided into groups of three in thousands, with "," as the thousands separator. For example, 331,449,281. Thousand separators are also widely used in Chinese, instead of introducing the ten thousand separator for 万. K is often used as a shorthand for thousands, and 5000 is written as 5K.

Table 2. *Decimal units in Chinese and English* **表2.**中文英语中的十进制单位

1	个	One	1
10	十	Ten	10
10^2	百	Hundred	100
10^3	千	Thousand	1,000
10^4	万	Ten Thousand	10,000
10^5	十万	Hundred Thousand	100,000
10^6	百万	Million	1,000,000
10^7	千万	Ten Million	10,000,000
10^8	亿	Hundred Million	100,000,000
10^9	十亿	Billion	1,000,000,000
10^{10}	百亿	Ten Billion	10,000,000,000
10^{11}	千亿	Hundred Billion	100,000,000,000
10^{12}	兆	Trillion	1,000,000,000,000

另一个我们要注意的是英语 11 到 19 有特别的单词（表 3），比如 15 不是 ten five，而是 fifteen。同样 20，30，到 90 也有专用词，比如 40 不是 four ten，而是 forty。有没有注意到 13-19 的英文词尾都是 teen，这也是 teenage 这个词的由来，指年龄在 13 到 19 岁的年轻人。

Another thing we should know is that there are special words for 11 to 19 (Table 3), for example, 15 is not ten-five, but fifteen. And there are also special words for 20, 30, up to 90. For example, 40 is not four-ten, but forty. You might have noticed that the endings of the words from 13 to 19 have the same teen at the end. This is the origin of the word teenage, which refers to young people between the ages of 13 and 19.

Table 3. *11-19 20-90 in English and Chinese* **表3.**中英文 *11–19 20-90*

11	Eleven	十一			
12	Twelve	十二	20	Twenty	二十
13	Thirteen	十三	30	Thirty	三十
14	Fourteen	十四	40	Forty	四十
15	Fifteen	十五	50	Fifty	五十
16	Sixteen	十六	60	Sixty	六十
17	Seventeen	十七	70	Seventy	七十
18	Eighteen	十八	80	Eighty	八十
19	Nineteen	十九	90	Ninety	九十

一万四千八百五十三

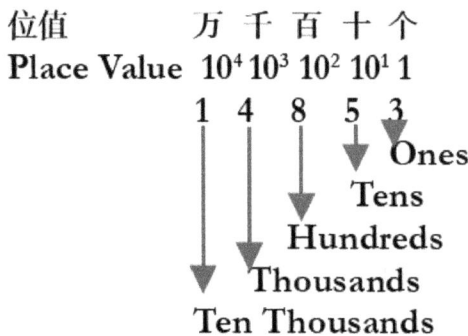

位值　　　万 千 百 十 个
Place Value 10^4 10^3 10^2 10^1 1
　　　　　1　4　8　5　3
　　　　　　　　　　　Ones
　　　　　　　　　　Tens
　　　　　　　　　Hundreds
　　　　　　　　Thousands
　　　　　　Ten Thousands

Fourteen thousand eight hundred fifty three

Figure 2. *Chinese English and Arabic number* **图2.** 中英和阿拉伯数字

从上面的例子（图 2）我们可以看出，中文的一万四千八百五十三，英文的 fourteen thousand eight hundred fifty three，和 14853，表示同一个数字，但阿拉伯数字 14853 最简洁。当我们学习运算时，基于阿拉伯数字的系统更是优越。阿拉伯数字，十进制和位值的完美结合，创造了一个简单易学并且高效的系统。

As we can see from the above example (Figure 2), the Chinese 一万四千八百五十三, the English fourteen thousand eight hundred fifty-three, and 14853, represent the same number, but the Arabic number 14853 is the most concise. When we learn arithmetic, the system based on Arabic numerals is even better. The perfect combination of Arabic numbers, decimal and place value creates a system that is easy to learn and efficient.

二进制 Binary

位值系统不仅适用十进制，理论上可以选择任何基数，假如基数不是 10 而是 2，那我们得到就是二进制（图 3）。基数为 2 的二进制是计算机里最基本的概念，今天世界上绝大部分运算实际上都是通过二进制完成的。

The place value system does not only apply to decimal, but theoretically any base can be chosen; if the base is not 10 but 2, then we get binary (Figure 3). Base-2 binary is the most basic concept in computers, and the vast majority of computations in the world today are actually done in binary.

$$\text{Place Value 位值} \quad 2^4\, 2^3\, 2^2\, 2^1\, 1$$
$$\text{Binary 二进制} \quad 1\ 0\ 0\ 1\ 1$$
$$1*2^4+0*2^3+0*2^2+1*2^1+1*1 = 19$$

Figure 3. *Binary place value* **图 3.** *二进制位值*

时间 Time

　　人类对时间的认知是一个漫长的过程，也因为此，时间的单位各种各样，非常复杂（表4）。时间的转换计算也常常被用作算术的练习。在科学和工程中，秒被定义为最基本的时间单位，十进制的运用更为广泛。

The human understanding of time is a long process, and because of this, the units of time are various and very complex (Table 4). Time conversion calculation is also often used as an exercise in arithmetic. In science and engineering, the second is defined as the most basic unit of time, and the decimal system is more widely used.

Table 4. *Units in Time* **表4.** *时间单位*

单位	Unit	转换	Conversion
秒; 秒钟	Second		
分; 分钟	Minute	60 秒	60 seconds
时; 小时	Hour	60 分	60 minutes
日; 天	Day	24 时	24 hours
周	Week	7 天	7 days
月	Month	30.43 天	30.43 days
季度	Season; quarter	3 月	3 months
年	Year	12 月	12 months
十年	Decade	10 年	10 years
世纪	Century	100 年	100 years
千年	Millennium	1000 年	1000 years

3 算术运算Arithmetic Operations

每个人都很熟悉加减乘除四种运算，又称四则运算，因为它们在日常生活中很常用。有时它们也被称为加法减法乘法和除法。运算的符号又叫运算符，+−*/，也很常见。+−*/作为运算时读作加减乘除，有时人们也会特别注明它们是符号，比如说+是加号，−是减号等等。其他的运算如取模（%）和乘方（**），大部分人只有在学习和考试中才可能见到。

Everyone is familiar with the four operations of addition, subtraction, multiplication and division, also known as the four basic operations in Math, because they are commonly used in everyday life. The symbols for the operations, also called operators, +−*/, are also common. +−*/ is read as addition, subtraction, multiplication, and division when used as an operation, and sometimes people make a point of stating that they are symbols, such as + is a plus sign, − is a minus sign, and so on. Other operations such as modulo (%) and exponentiation (**) are only likely to be seen by most people when studying Math and taking tests.

加 Addition

加法是求两个数的和，如下图中的例子（图4），17 + 8 = 25，读作十七加八等于二十五。

Addition is finding the sum of two numbers, as in the example below (Figure 4), 17 + 8 = 25, which reads seventeen plus eight equals twenty-five.

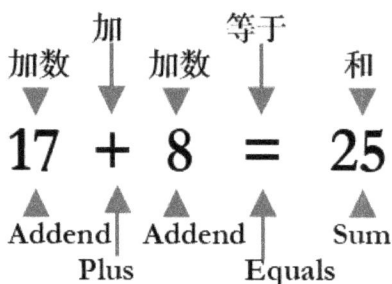

Figure 4. Addition 图4. 加法

加法满足交换律和结合律。

交换律是指两个加数在前和在后结果是相同的，即 17 + 8 = 8 + 17。

结合律是说如果有多个数相加，运算的顺序是无关紧要的。例如：

17 + 8 + 3 = (17 + 8) + 3 = 17 + (8 + 3)。

这里我们用括号( ) 表示括号里面的运算先一步进行。也就是说我们可以先做 17 + 8 再加 3，也可以先做 8 + 3 再加 17。结合前面的交换律，我们还可以先做 17 + 3 再加 8。

Addition has the commutative property and the associative property.

The commutative property means that the sum is the same when the two terms in addition exchange order, for example, $17 + 8 = 8 + 17$.

The associative property means that when three or more numbers are added together, the order of operations is irrelevant. For example:

$$17 + 8 + 3 = (17 + 8) + 3 = 17 + (8 + 3).$$

Here we use parentheses ( ) to indicate that the operation inside the parentheses is done first. This means that we can do $17 + 8$ and then add 3, or we can do $8 + 3$ and then add to 17. Combined with the commutative property, we can also do $17 + 3$ first and then add 8.

两数相加最简洁的方法是将加数分两行按位值对齐（图 5），个位对个位，十位对十位，然后从个位加起，再十位，再更高位。如果在一个数位上两数相加结果大于 9，就写下和数的个位，而将十位的 1 加在下一个高一位的数位。这个过程叫进位（图 5），通常是在进位的顶部或者底部另写上一个 1 字，以表示这里有进位。

The simplest way to add two numbers is to arrange the addends into two rows (Figure 5) and align them by place value, ones place to ones place, tens place to tens place, and then add from the ones place, then tens place, and then higher places. If the sum of two numbers in one digit is greater than 9, write the ones digit of the sum and add the 1 in the tens digit to the next higher digit. This process is called carrying (Figure 5), and usually an extra 1 is written at the top or bottom of the digit to indicate that there is a carry.

Carry 进位

▼

$$
\begin{array}{r}
\overline{1\ 1\ 1} \\
1\ 7\ 7\ 6 \\
+\quad 2\ 4\ 8 \\
\hline
2\ 0\ 2\ 4
\end{array}
$$

Figure 5. Carry in addition 图 *5.* 加法进位

减 Subtraction

减法是求两个数的差，从被减数中减去减数，可以看成是加法的逆运算。当我们需要验证减法的运算结果时，可以反过来用加法，差加上减数应该等于被减数。下图中的例子（图6）25 - 8 = 17 读作二十五减八等于十七。

Subtraction is to find the difference between two numbers. Subtracting the subtrahend (right-hand side) from the minuend (left-hand side) can be regarded as the inverse operation of addition. When we need to verify the result of subtraction, we can use addition in reverse. The difference plus the subtrahend should equal the minuend. The example 25 − 8 = 17 in the figure below (Figure 6) is read as twenty-five minus eight equals seventeen.

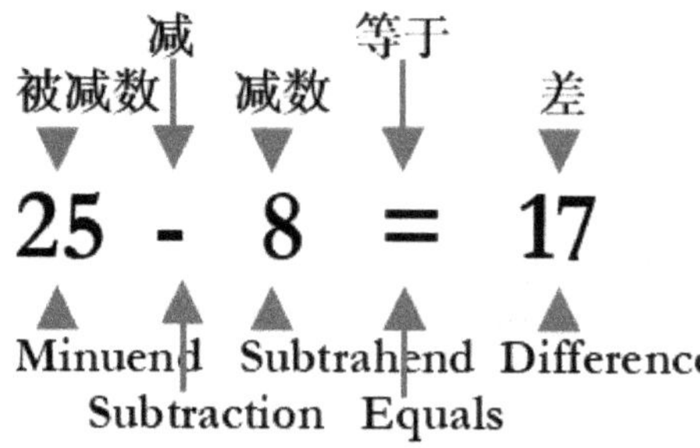

Figure 6. Subtraction **图 6.** 减法

多位数的减法算法与加法相似，将第一项被减数与第二项减数按位值对齐，然后从个位开始计算（图7）。如果一个数位上的被减数小于减数，必须向高一位借1相当于10，这种方法叫退位或者借位。退位可以将数字明确写出来，也可以简化用一个退位点表示（图7）。

The multi-digit subtraction algorithm is similar to addition. The first term minuend and the second term subtrahend are aligned according to their place value, and then the calculation starts from the ones digit

(Figure 7). If the minuend of a digit is smaller than the subtrahend, 1 must be borrowed from the higher digit, which is equivalent to 10. This method is called regrouping or borrowing. The borrowing can be written clearly as a number, or it can be simplified and represented by a dot (Figure 7).

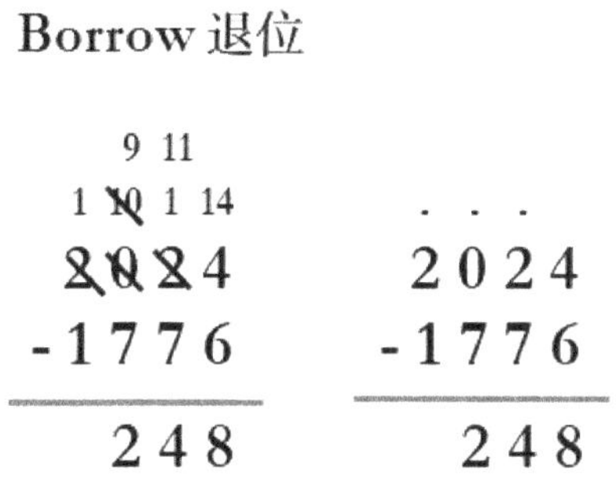

Borrow 退位

```
    9  11
  1 10 1 14
   2 0 2 4          2 0 2 4
  -1 7 7 6         -1 7 7 6
  ─────────        ─────────
     2 4 8            2 4 8
```

Figure 7. Borrow in subtraction 图 7. 减法退位

负数和绝对值 Negative Number and Absolute Value

当被减数小于减数时，两者的差小于 0，叫做负数。负数用负号 − 表示，如 −2 是负二。

有了负数，减法也可以看成是加法的一种。如 5 − 3 = 5 + (−3)，即减法是加负数。

自然数是非负整数，即包含 0 和所有正整数。

负有时也会被看成是一种运算，即在一个数前加上一个负号。负数的负数是正数。

与负数相对应的是正数，其数值为正，有时也加上"+"号以示强调，如 +5。

一个数的绝对值是指去掉正负号后的非负数值，即正数 x 的绝对值是其本身，负数 x 的绝对值是 $-x$，0 的绝对值是 0。一个数的绝对值也可看作是它和零之间的距离。

When the minuend is smaller than the subtrahend, the difference between the two is less than 0, and it is called a negative number. Negative numbers are represented by the negative sign $-$, for example -2 is negative two.

With negative numbers, subtraction can also be considered a type of addition. For example, $5 - 3 = 5 + (-3)$, that is, subtraction is addition of negative numbers.

Natural numbers are non-negative integers, which include 0 and all positive integers.

Negative is sometimes seen as an operation, which involves adding a negative sign in front of a number. The negative of a negative number is a positive number.

Corresponding to negative number is positive number, whose numerical value is positive, and sometimes a "+" sign is added for emphasis, such as $+5$.

The absolute value of a number refers to the non-negative value after removing the $+/-$ sign, that is, the absolute value of a positive number x is itself, the absolute value of a negative number x is $-x$, and the absolute value of 0 is 0. The absolute value of a number can also be thought of as the distance between it and zero.

乘 Multiplication

将一个数多次相加，就得到乘法。比如 3 + 3 + 3 + 3，4 个 3 相加，等同于 3 × 4。× 是乘号，有时也以 * 或者 · 代替。3 × 4 读作三乘（以）四。相乘的两个数叫乘数，也叫因子，结果叫积（图 8）。

When you add a number multiple times, you get multiplication. For example, 3 + 3 + 3 + 3, the sum of four 3's is equivalent to 3 × 4. × is a multiplication sign, sometimes it is replaced by * or · symbols. 3 × 4 is pronounced as three multiplied by four. The two numbers multiplied together are called multipliers, also called factors, and the result is called the product (Figure 8).

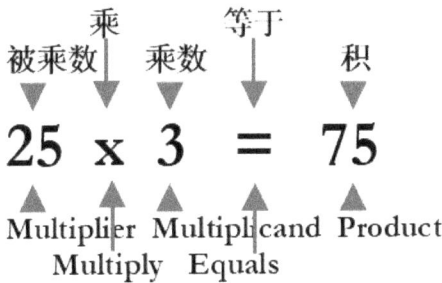

$$\underset{\text{Multiplier}}{\overset{\text{被乘数}}{25}} \; \underset{\text{Multiply}}{\overset{\text{乘}}{\times}} \; \underset{\text{Multiplicand}}{\overset{\text{乘数}}{3}} \; \underset{\text{Equals}}{\overset{\text{等于}}{=}} \; \underset{\text{Product}}{\overset{\text{积}}{75}}$$

Figure 8. *Multiplication* **图 8.** *乘法*

乘法满足交换律，结合律，和分配律。

交换律：3 × 4 = 4 × 3

结合律：(3 × 4) × 5 = 3 × (4 × 5)

分配律：3 × (4 + 5) = 3 × 4 + 3 × 5

Multiplication satisfies the commutative, associative, and distributive laws.

Commutative law: 3 × 4 = 4 × 3

Associative law: (3 × 4) × 5 = 3 × (4 × 5)

Distributive law: 3 × (4 + 5) = 3 × 4 + 3 × 5

乘法表 Multiplication Table

中国的小学生必须熟记九以内的乘法，也就是必须记住九九乘法口诀表（表5）。在英美等学校，学生要求熟悉乘法表，通常是 12 × 12 的乘法表（表6）。

Chinese pupils must memorize multiplication within nine, which means they must memorize the nine-nine multiplication table（Table 5). In Western schools, students are required to be familiar with the multiplication table, usually a 12 × 12 multiplication table (Table 6).

Table 5. *Nine-nine Multiplication Table* **表5.**九九乘法口诀表

一一 得一 1x1=1								
一二 得二 1x2=2	二二得 四 2x2=4							
一三 得三 1x3=3	二三得 六 2x3=6	三三得 九 3x3=9						
一四 得四 1x4=4	二四得 八 2x4=8	三四一 十二 3x4=12	四四一 十六 4x4=16					
一五 得五 1x5=5	二五一 十 2x5=10	三五一 十五 3x5=15	四五二 十 4x5=20	五五二 十五 5x5=25				
一六 得六 1x6=6	二六十 一二 2x6=12	三六一 十八 3x6=18	四六二 十四 4x6=24	五六三 十 5x6=30	六六三 十六 6x6=36			
一七 得七 1x7=7	二七一 十四 2x7=14	三七二 十一 3x7=21	四七二 十八 4x7=28	五七三 十五 5x7=35	六七四 十二 6x7=42	七七四 十九 7x7=49		
一八 得八 1x8=8	二八一 十六 2x8=16	三八二 十四 3x8=24	四八三 十二 4x8=32	五八四 十 5x8=40	六八四 十八 6x8=48	七八五 十六 7x8=56	八八六 十四 8x8=64	
一九 得九 1x9=9	二九一 十八 2x9=18	三九二 十七 3x9=27	四九三 十六 4x9=36	五九四 十五 5x9=45	六九五 十四 6x9=54	七九六 十三 7x9=63	八九七 十二 8x9=72	九九八 十一 9x9=81

Table 6. *12 × 12 Multiplication Chart* **表6.** 十二乘十二乘法表

×	1	2	3	4	5	6	7	8	9	10	11	12
1	1	2	3	4	5	6	7	8	9	10	11	12
2	2	4	6	8	10	12	14	16	18	20	22	24
3	3	6	9	12	15	18	21	24	27	30	33	36
4	4	8	12	16	20	24	28	32	36	40	44	48
5	5	10	15	20	25	30	35	40	45	50	55	60
6	6	12	18	24	30	36	42	48	54	60	66	72
7	7	14	21	28	35	42	49	56	63	70	77	84
8	8	16	24	32	40	48	56	64	72	80	88	96
9	9	18	27	36	45	54	63	72	81	90	99	108
10	10	20	30	40	50	60	70	80	90	100	110	120
11	11	22	33	44	55	66	77	88	99	110	121	132
12	12	24	36	48	60	72	84	96	108	120	132	144

除 Division

　　除法是乘法的逆运算，算符常见有 ÷ 和 / 。但在小学的初始阶段，学的是自然数的除法，可以看成是重复的减法。比如 11 ÷ 4 相当于 11 − 4 − 4，直到余数为 3，不能再减 4，否则余数为负数。也就是说 11 ÷ 4 = 2 余数 3（图 9）。除法的结果称为商。

Division is the inverse operation of multiplication, and common operators are ÷ and /. But in the initial stage of elementary school, what is learned is the division of natural numbers, which can be regarded as repeated subtraction. For example, 11 ÷ 4 is equivalent to 11 − 4 − 4. Until the remainder is 3, 4 cannot be subtracted from 3, otherwise the remainder will be a negative number. That is, 11 ÷ 4 = 2 remainder 3 (Figure 9). The result of division is called quotient.

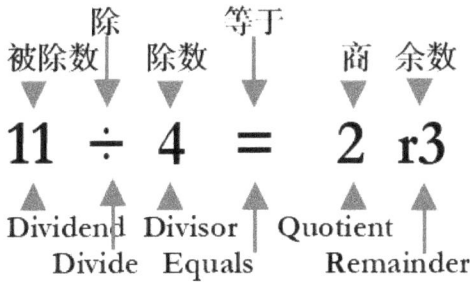

Figure 9. Division **图 9.** *除法*

除法的计算一般用长除法（又叫直除法）的方法（图10）。

The calculation of division generally uses the method of long division (Figure 10).

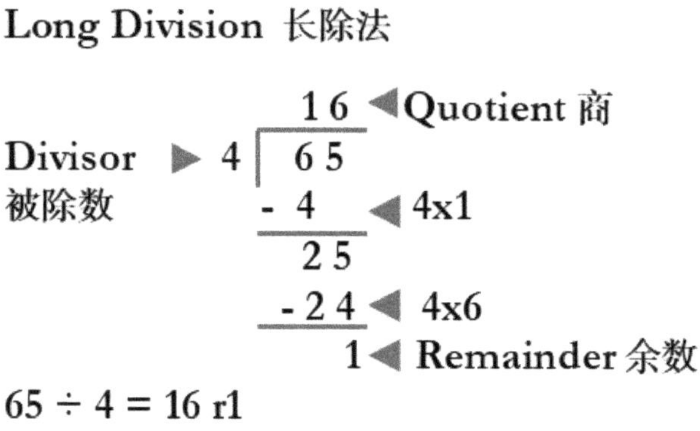

Long Division 长除法

```
                16  ◀ Quotient 商
Divisor  ▶  4 ⟌ 6 5
被除数          - 4      ◀ 4x1
               ‾‾‾
                2 5
              - 2 4  ◀ 4x6
               ‾‾‾
                  1  ◀ Remainder 余数
      65 ÷ 4 = 16 r1
```

Figure 10. Long division 图 10. 长除法

能被 2 整除的自然数叫偶数，否则叫奇数，0 是偶数。自然数要么是偶数要么是奇数，这种属性叫奇偶性。偶数又叫双数，奇数又叫单数。

除法的除数不能为 0，比如 5/0 的结果是无穷大，而 0/0 的结果是不确定的。

A natural number that is divisible by 2 is called an even number, otherwise it is called an odd number, and 0 is an even number. Natural numbers are either even or odd. This property is called parity. Even numbers are also called double numbers.

The divisor of division cannot be 0. For example, the result of 5/0 is infinity, but the result of 0/0 is uncertain.

单位元 Identity Element

单位元也称恒等元，是指运算时，如果一项为单位元，运算结果会与另一项相同。从下面的例子可以看出，0 是加法和减法的单位元，1 是乘法和除法的单位元。

The identity element is also called the neutral element, which means that if one term is the identity element in a calculation, the result of the operation will be the same as the other term. As can be seen from the example below, 0 is the identity element for addition and subtraction, and 1 is the identity element for multiplication and division.

$25 + 0 = 25$

$25 - 0 = 25$

$25 * 1 = 25$

$25 \div 1 = 25$

取模 Modulo

取模又称模除，是求两个正整数相除的余数，运算符是 % 或者 mod。如 9 = 2 * 4 + 1，9 除 4 的结果不是一个整数，会有余数 1。用取模运算表示，得到的是 9 % 4 = 1。取模在计算机编程中有着广泛的应用。

Modulo, also known as modular division, is to find the remainder of the division of two positive integers. The operator is % or mod. For example, $9 = 2 * 4 + 1$, the result of dividing 9 by 4 is not an integer, there will be a remainder of 1. Expressed using the modulo operation, the result is $9 \% 4 = 1$. Modulo operation is widely used in computer programming.

乘方 Exponentiation

相同的数重复相乘的运算叫乘方，其结果叫做幂（图 11）。乘方的运算符号常用的有 ^ 和 **，更多时用上标表示。

The operation of repeated multiplication of the same number (base) is called exponentiation, and the result is called power (Figure 11). Commonly used symbols for exponentiation operation include ^ and **, and more often superscripts.

乘方 Exponential

$$5*5*5 = 5\wedge 3 = 5**3 = 5^3$$

底数 Base ▶5^3◀ Exponent 指数

Power 幂

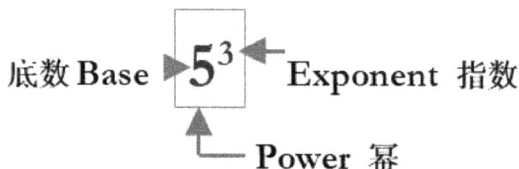

Figure 11. Exponential **图 11.** 乘方

5^3 读作五的三次方。如果指数是 2，则称为平方；指数为 3，又称为立方。

5^3 is read as five (raised) to the third power. If the exponent is 2, it is called a square; if the exponent is 3, it is also called a cube.

运算的次序 Order of Operations

运算的次序遵守以下规律：

（1）括号，如有多重括号，里面的括号优先。

（2）乘方

（3）乘和除

（4）加和减

（5）从左到右

The order of operations follows the following rules:
(1) Parentheses. If there are multiple parentheses, the parentheses inside take precedence.
(2) Exponentiation
(3) Multiplication and division
(4) Addition and subtraction
(5) From left to right

18 - 72 / (3 + (4 + 5)) * 2 Inside parenthesis 内括号

= 18 - 72 / (3 + 9) * 2   Parenthesis 括号

= 18 - 72 / 12 * 2 Multiplication and divide, left to right
                 乘除，从左到右

= 18 - 6 * 2   乘除

= 18 − 12

= 6

Figure 12. Order of operations **图 12.** 运算的次序

27

有时一些运算的表达容易产生歧义，比如 16÷2(2+2)，究竟是(16÷2)*(2+2)还是16÷[2*(2+2)]？类似a/b/c可以是(a/b)/c，也可能是a/(b/c)。书写时应该避免歧义，以免引起误解。

Sometimes there may be ambiguity in the expression of some operations, such as 16÷2(2+2). Is it (16÷2)*(2+2) or 16÷[2*(2+2)]? Similarly a/b/c could be (a/b)/c or it could be a/(b/c). Ambiguity should be avoided when writing to avoid misunderstandings.

比较 Comparison

日常生活中我们常常要比较两个数的大小，如 5 > 3，读作 5 大于 3。

In daily life, we often need to compare the size of two numbers, such as 5 > 3, which is pronounced as 5 greater than 3.

Table 7. Comparison 表7. 比较

符号 Symbol	例子 Example	English	中文
=	$5 - 2 = 3$	equal	等于
≠	$5 \neq 3$	not equal	不等于
>	$5 > 3$	greater than	大于
<	$3 < 5$	less than	小于
≥ or >=	$x \geq 3$	not less than, greater or equal, at least	不小于，大于等于
≤ or <=	$y \leq 3$	not greater than, less or equal, at most	不大于，小于等于

4 分数 Fraction

分数是指将整数分成等份，如 4/7 是 4 被 7 等分（图 13）。1/2 读作二分之一，4/7 读作七分之四。分数也可以被看成整数的除法，或者两个数的比值。分数值小于 1，即分子小于分母的分数，叫真分数。与真分数相对的是假分数（图 14），其分子大于或者等于分母。假分数可以变成一个整数部分外加分数部分，如 $17/5 = 3\frac{2}{5}$，这种形式叫带分数。$3\frac{2}{5}$ 读作三又五分之二。

A fraction refers to dividing a whole number into equal parts, for example, 4/7 is 4 divided into 7 equal parts (Figure 13). 1/2 is pronounced as one-half, and 4/7 is pronounced as four-sevenths. Fractions can also be thought of as the division of whole numbers, or the ratio of two numbers. A fraction whose value is less than 1, that is, a fraction whose numerator is smaller than its denominator, is called a proper fraction. The opposite of a proper fraction is an improper fraction (Figure 14), where the numerator is greater than or equal to the denominator. An improper fraction can be turned into an integer part plus a fraction part, such as $17/5 = 3\frac{2}{5}$. This form is called a mixed number. $3\frac{2}{5}$ is pronounced three and two-fifths.

分数 Fraction

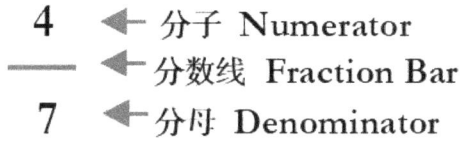

$$\frac{4}{7} \begin{array}{l} \leftarrow \text{分子 Numerator} \\ \leftarrow \text{分数线 Fraction Bar} \\ \leftarrow \text{分母 Denominator} \end{array}$$

Figure 13. *Fraction* **图 13.** 分数

假分数 Improper Fraction

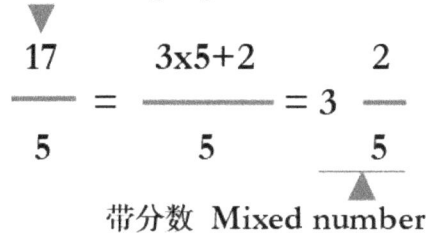

$$\frac{17}{5} = \frac{3 \times 5 + 2}{5} = 3\frac{2}{5}$$

带分数 Mixed number

Figure 14. *Improper fraction and mixed fraction* **图 14.** 假分数和带分数

分数的乘法和除法 Multiplication and Division of Fractions

当两个分数相乘时，将它们的分子和分母分别相乘。当两个分数相除时，如果我们将除数的分子分母互换，除法将变成乘法（图15）。

When multiplying two fractions, multiply their numerators and denominators separately. When dividing two fractions, if we swap the numerator and denominator of the divisor, the division becomes multiplication (Figure 15).

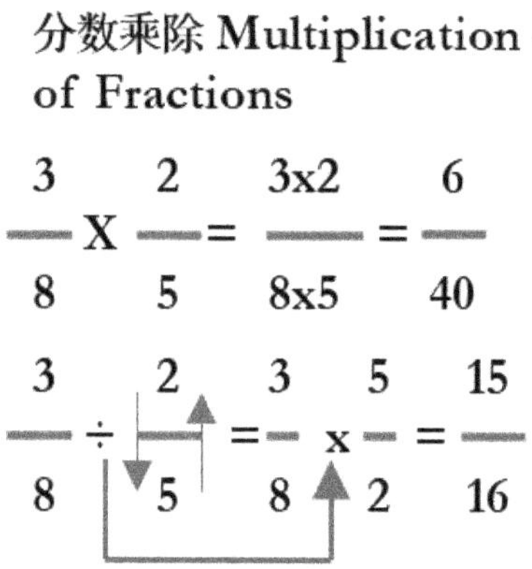

分数乘除 Multiplication & Division of Fractions

$$\frac{3}{8} \times \frac{2}{5} = \frac{3 \times 2}{8 \times 5} = \frac{6}{40}$$

$$\frac{3}{8} \div \frac{2}{5} = \frac{3}{8} \times \frac{5}{2} = \frac{15}{16}$$

Figure 15. Multiplication and division of fractions 图 15. 分数乘除

我们前面说过乘法和除法的单位元是1。1也可以写成分数形式，分子分母相同，如5/5。任何一个分数乘1数值不变，也就是说如果分子分母同时乘一个数（非零），数值不变。例如，1/2 * 1 = 1/2 * 5/5 = 5/10。类似也可以分子分母同时除一个数，如6/40 =（6÷2）/（40÷2）= 3/20。

We said earlier that the identity element for multiplication and division is 1. 1 can also be written as a fraction, with the same numerator and denominator, such as 5/5. The value of any fraction multiplied by 1 does not change. That is to say, if the numerator and denominator are multiplied by a number (non-zero) at the same time, the value does not change. For example, $1/2 * 1 = 1/2 * 5/5 = 5/10$. Similarly, you can also divide the numerator and denominator at the same time by the same number, such as $6/40 = (6÷2)/(40÷2) = 3/20$.

整数分解 Integer Factorization

整数分解又叫因数分解或者因子分解，将一个整数分解成更小的整数（因子）的乘积，通常我们还会要求这里的因子都是正整数。例如75可以分解成5 × 15，3 × 25，或者5 × 5 × 3（图16）。

Integer decomposition is also called factorization or factoring. It decomposes an integer into the product of smaller integers (factors), usually with the requirement that the factors are positive. For example, 75 can be broken down into 5 × 15, 3 × 25, or 5 × 5 × 3 (Figure 16).

整数分解 Integer Factorization

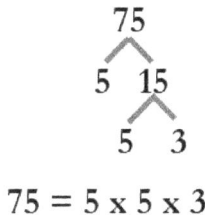

$$75 = 5 \times 5 \times 3$$

Figure 16. Integer factorization **图 16.** 整数分解

质数分解 Prime Decomposition

质数又称素数，只可以被1和自己整除。按定义，1不是质数。例如7是质数，60不是质数。不是质数的自然数又被称为合数。

33

任何一个大于 1 的自然数, 如果自身不为质数, 那么可以唯一分解成有限个质数的乘积。这个定理被称为算术基本定理。以 60 为例, 其整数分解并不唯一, 可以是 5 × 12, 也可以是 6 × 10。但它有唯一的质数分解: $60 = 2 \times 2 \times 3 \times 5$。

A prime number, or a prime, is only divisible by 1 and itself. By definition, 1 is not a prime number. For example, 7 is a prime number, but 60 is not a prime number. Natural numbers that are not prime numbers are also known as composite numbers.

Any natural number greater than 1, if it is not a prime number, can be uniquely decomposed into the product of a finite number of prime numbers. This theorem is called the Fundamental Theorem of Arithmetic. Taking 60 as an example, its integer decomposition is not unique, it can be 5 × 12 or 6 × 10. But it has a unique prime factorization: $60 = 2 \times 2 \times 3 \times 5$.

分数加减 Addition and Subtraction of Fractions

如果两个分数的分母相同, 两个分数的加减就是两个分子的加减, 分母不变。当两个加数分母不相同时, 必须首先将分母变成相同。

If the denominators of two fractions are the same, the addition and subtraction of the two fractions is the addition and subtraction of the two numerators, leaving the denominator unchanged. When the denominators of two addends are not the same, the denominators must first be made the same.

最小公分母Least Common Denominator

当两个或多个分数具有相同的分母时，它们有公共分母，简称公分母。只有当两个分数具有公分母时，我们才可以做分数的加法和减法。

如果两个分数不具有相同的分母，我们可以通过将分子分母同时乘以一个量，既保持分数值不变，又改变分母的大小（图 17）。处理得当，我们可以找到两个分数的公分母，这个过程叫通分。分子分母同时除以一个量叫约分。

When two or more fractions have the same denominator, they have a common denominator. We can add and subtract fractions only if the two fractions have a common denominator.

If two fractions do not have the same denominator, we can keep the value of the fraction constant while changing the size of the denominator by multiplying both the numerator and the denominator by the same quantity (Figure 17). Done properly, we can find the common denominator of two fractions. Dividing the numerator and denominator simultaneously by the same quantity is called reduction.

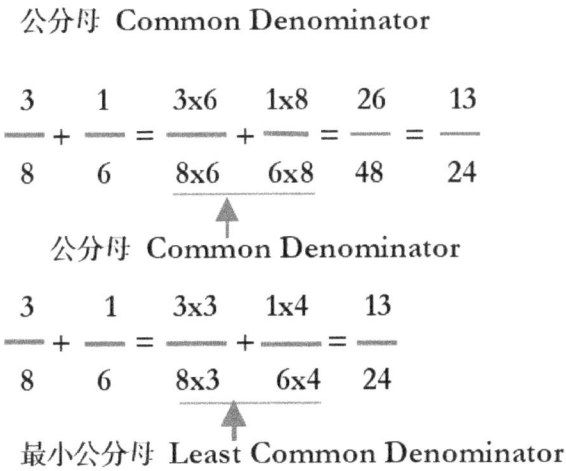

公分母 Common Denominator

$$\frac{3}{8} + \frac{1}{6} = \frac{3 \times 6}{8 \times 6} + \frac{1 \times 8}{6 \times 8} = \frac{26}{48} = \frac{13}{24}$$

公分母 Common Denominator

$$\frac{3}{8} + \frac{1}{6} = \frac{3 \times 3}{8 \times 3} + \frac{1 \times 4}{6 \times 4} = \frac{13}{24}$$

最小公分母 Least Common Denominator

Figure 17. *Least common denominator* ***图 17.*** *最小公分母*

最小公分母是最小公倍数的一个特列，唯一的区别是最小公分母的数字是分母。两个整数的最小公倍数是可以被两个整数整除的最小整数。

寻找最小公倍数的简单方法是分别计算两数的倍数，然后从中找到最小的相同结果（表8）。在下面的表格中，我们寻找8和6的最小公倍数。首先我们分别计算8和6的倍数，然后在它们的倍数中寻找第一个共同的数。很快我们就会发现24是8的倍数也是6的倍数，这就是我们要找到最小公倍数。

The least common denominator (LCD) is a special case of the least common multiple (LCM). The only difference is that the number in the least common denominator is the denominator. The least common multiple of two integers is the smallest integer that is divisible by the two integers.

A simple way to find the least common multiple is to calculate the multiples of two numbers separately and then find the smallest identical result from them (Table 8). In the table below, we find the least common multiple of 8 and 6. First we calculate the multiples of 8 and 6 respectively, and then find the first common number among their multiples. Soon we will find that 24 is a multiple of 8 and a multiple of 6, which is the least common multiple we try to find.

Table 8. *Find least common denominator* **表8.** 寻找最小公倍数

	8	6
1	8	6
2	16	12
3	24	18
4	32	24
5	40	30

另一种方法是将每个分母分解成质数的积。

在上面的例子里，8 和 6 会被分解成如下：

8 = 2 * 2 * 2

6 = 2 * 3

确定分母中每个质数出现的最大次数，如 2 在 8 中出现 3 次，3 在 6 中出现 1 次。将每个质数，按出现的最大次数，相乘，就得到最小公分母 2 * 2 * 2 * 3 = 24。

Another way is to factor each denominator into the product of prime numbers.

In the above example, 8 and 6 would be broken down as follows:

8 = 2 * 2 * 2

6 = 2 * 3

Determine the maximum number of occurrences of each prime number in the denominator, such as 2 appearing 3 times in 8 and 3 appearing once in 6. Multiply each prime number together according to the maximum number of occurrences to get the lowest common denominator 2 * 2 * 2 * 3 = 24.

与最小公倍数相对应的还有最大公约数（最大公因数）。一个整数如是另一个整数的倍数，后者称为前者的约数或者因子。最大公约数是指一组数中共同的约数中的最大的那个。8 和 6 的最大公约数是 2；24 和 18 的最大公约数是 6。利用最大公约数我们也能得到最小公倍数（图 18）。

Corresponding to the least common multiple is the greatest common divisor (GCD, greatest common factor). If an integer is a multiple of another integer, the latter is called a divisor or factor of the previous one. The greatest common divisor refers to the largest common divisor in a set of numbers. The greatest common divisor of 8 and 6 is 2; the greatest common divisor of 24 and 18 is 6. Using the greatest common divisor we can also find the least common multiple (Figure 18).

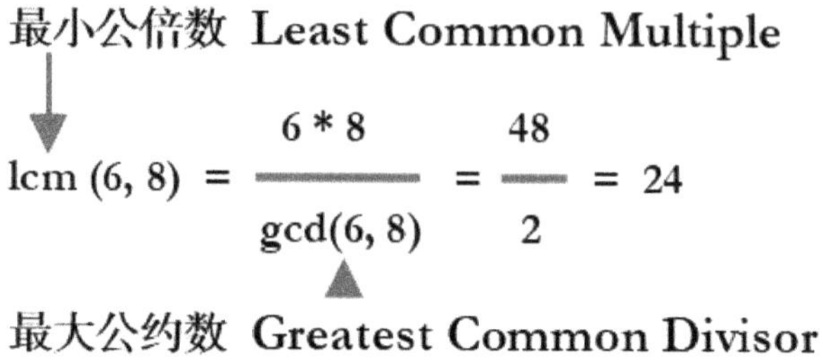

最小公倍数 Least Common Multiple

$$\text{lcm}(6, 8) = \frac{6 * 8}{\gcd(6, 8)} = \frac{48}{2} = 24$$

最大公约数 Greatest Common Divisor

Figure 18. *Least common multiple and greatest common divisor* **图 18.** *最小公倍数和最大公约数*

百分比Percentage

百分比是分数的一种，分母为100，用百分号%表示。如35%代表35/100，读作百分之三十五。

中文中常用成和折表示百分之十，比如80%有时会说成八成或者八折，常用于减价促销。

Percentage is a type of fraction with a denominator of 100 and is represented by the percent sign %. For example, 35% means 35/100, which is read as thirty-five percent.

In Chinese 成 and 折 are commonly used to indicate ten percent, for example, 80% is sometimes said to be 8 成 or 8 折, often used to advertise price cuts and promotions.

倒数Reciprocal

如果两个数相乘的积为1，那么这两个数互为倒数。一个数的倒数是1除以这个数。一个分数的倒数是将其分子分母互换，如3/4的倒数是4/3，同样4/3的倒数是3/4。0没有倒数。倒数也称为乘法逆元。

If the product of multiplying two numbers is 1, then those two numbers are the reciprocal of each other. The reciprocal of a number is 1 divided by that number. The reciprocal of a fraction is to swap its numerator and denominator, e.g., the reciprocal of 3/4 is 4/3, likewise the reciprocal of 4/3 is 3/4. 0 has no reciprocal. Reciprocal is also called multiplicative inverse.

小数 Decimal Representation

小数是整数十进制的进一步延伸，用以表达 10 的分数（图 19）。通常我们用一个点来标识点之后的位值是十分之一，百分之一，千分之一，等等。小数点又叫小数分隔号，在欧洲会用逗号"，"代替。小数点读作点，如 12.45 读作十二点四五。

Decimals are a further extension of the integer decimal system and are used to express fractions of 10 (Figure 19). Usually we use a dot to indicate that the place value after the dot is one-tenth, one-hundredth, one-thousandth, etc. The decimal point is also called the decimal separator, which is replaced by a comma "," in Europe. The decimal point is read as point, for example, 12.45 is read as twelve point four five.

小数 Decimal Representation

Place Value
位值   10 1 10^{-1} 10^{-2}
                1 2 . 4 5
             ▲
Decimal point 小数点

Figure 19. Decimal representation **图 19.** 小数表示法

如果一个小数小数点后面的位数是有限的，叫做有限小数。如果小数点后的位数是无限的，则叫做无限小数（图 20）。

无限小数又分为无限循环小数和无限不循环小数。无限循环小数的小数部分会出现无限重复的小数，而无限不循环小数的小数却没有有规律的重复。有限小数和无限循环小数都可以转化成分数，称为有理数。无限不循环小数不能转化成分数，称为无理数（图 20）。

If a decimal has a finite number of digits after the decimal point, it is called a finite decimal. If the number of digits after the decimal point is infinite, it is called an infinite decimal (Figure 20).

Infinite decimals are divided into infinite recurring decimals and infinite non-repeating decimals. The decimal part of an infinitely recurring decimal will have infinitely repeated decimals, while the decimals of an infinite non-repeating decimal will not repeat regularly. Both finite decimals and infinite recurring decimals can be converted into fractions, which are called rational numbers. Infinite non-repeating decimals cannot be converted into fractions and are called irrational numbers (Figure 20).

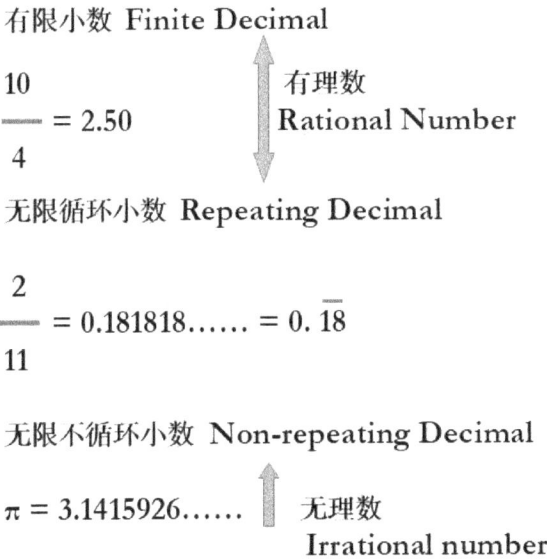

有限小数 Finite Decimal

$$\frac{10}{4} = 2.50 \qquad \begin{array}{l} \text{有理数} \\ \text{Rational Number} \end{array}$$

无限循环小数 Repeating Decimal

$$\frac{2}{11} = 0.181818\ldots\ldots = 0.\overline{18}$$

无限不循环小数 Non-repeating Decimal

$$\pi = 3.1415926\ldots\ldots \qquad \begin{array}{l} \text{无理数} \\ \text{Irrational number} \end{array}$$

Figure 20. *Rational and irrational number* **图 20.** *有理数和无理数*

小数单位 Prefix

Table 9. Prefix **表9.** 小数单位

Factor	Prefix	Symbol	中文
10^3	kilo-	k	千
10^{-1}	deci-	d	分
10^{-2}	centi-	c	厘
10^{-3}	milli-	m	毫
10^{-6}	micro-	μ	微
10^{-9}	nano-	n	纳/奈
10^{-12}	pico-	p	皮

舍入 Rounding

据统计 2020 年中国人口是 1,411,778,724，但由于统计的误差，这个数字并不是准确数字，我们通常只会用这个数字的前几位，简单地说 14.1 亿。同样美国人口 3.3 亿，也是一个简单且可以接受的近似。这种用简单的近似取代冗长的数字的方法，叫做舍入，又叫修约。

According to a survey, China's population in 2020 is 1,411,778,724. However, due to statistical errors, this number is not accurate. We usually only use the first few digits of this number, simply 1.41 billion. Similarly, to state the population of the United States is 330 million is also a simple and acceptable approximation. This method of replacing long numbers with simple approximations is called rounding.

常见的舍入是将数字缩短到特定的位值，比如钱币的数值会缩短到小数点后两位——￥10.2223 会近似为￥10.22。

不同的应用中有各种舍入的方法，下面介绍的是最常用的一种，简单地称为四舍五入，规则也很简单（图21）：

（1）四舍：在需要保留的位数后一位，如数值小于5，后面的尾数全部舍弃。

（2）五入：在需要保留的位数后一位，如数值不小于5，向前一位进一（后面的尾数同样全部舍弃）。

A common rounding is to shorten a number to a specific place value. For example, the value of money will be shortened to two decimal places, for example $10.2223 will be approximately $10.22.

There are various rounding methods in different applications. The following is the most commonly used one. The rules are also very simple (Figure 21):

(1) Round down: After the number of digits to be retained, if the value is less than 5, all subsequent digits will be discarded.

(2) Round up: After the number of digits to be retained, if the value is not less than 5, add one to the previous digit (all subsequent digits are also discarded).

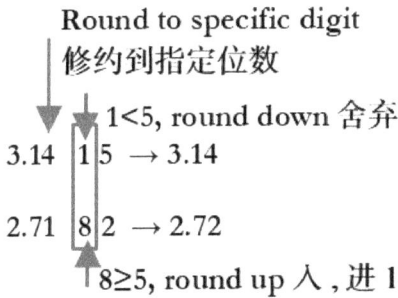

Round to specific digit
修约到指定位数

1<5, round down 舍弃
3.14 1|5 → 3.14

2.71 8|2 → 2.72
8≥5, round up 入，进1

Figure 21. *Round down and round up* **图 21.** 四舍五入

5　计算工具 Calculation Tool

自古以来人们就使用各种工具帮助计数和计算。还记得小时用手指算加减吗？历史上人们用过珠子、绳结、算筹（棍子）等等，而今天我们生活中离不开的手机和计算机，本质上都是超级计算工具。

Since ancient times people have used a variety of tools to help with counting and calculation. Remember using your fingers to add and subtract when you were a kid? Historically, people have used beads, knots, counting rods, and so on, and today we can't live without our cell phones and computers, all of which are essentially super computing tools.

算盘 Abacus

公元一千年后几经改进的算盘（图22）成为重要的计算工具，它的优势与阿拉伯数字类似，使用十进制的位值系统。算盘不仅可用做加减乘除运算，还可以用以做更复杂的运算。十七世纪一位中国王子，用 81 位 的 大 算 盘 ，成 功 地 计 算 了 $2^{1/12}$（ =*1.05946309435929526456182*5 ），精度达到25位有效数字。

After the first millennium A.D., the abacus (Figure 22), which was improved several times, became an important computational tool. Its advantages are similar to those of Arabic numerals, using a decimal place value system. The abacus could be used not only for addition, subtraction, multiplication and division, but also for more complex operations. A Chinese prince in the seventeenth century, using a large 81 position abacus, successfully calculated $2^{1/12}$ (=*1.059463094359295264561825*) with an accuracy of 25 significant figures.

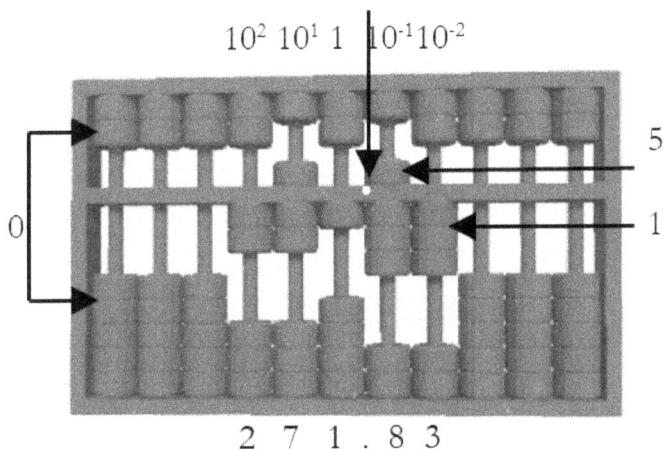

假想小数点 Imaginary Decimal Point

10^2 10^1 1 10^{-1} 10^{-2}

5

0

1

2 7 1 . 8 3

Figure 22. Abacus **图 22.** *算盘*

计算器 Calculator

电子计算器（图 23）的出现很快取代了机械的计算器如算盘，计算尺等。电子计算的能力远远超越任何机械计算。但计算不再直观，不再显示步奏，甚至不再用我们熟悉的十进制，而是转换成二进制。

因此有了全新的语言，叫做计算机语言，用来给计算机输入数据和指令，与其交流。控制计算机的指令集又叫软件，编写软件叫做编程。今天编程是每个工程师的必修课。手持计算器也是每个学生必备的工具。

The emergence of electronic calculators (Figure 23) quickly replaced mechanical calculators such as abacus, slide rule, etc. The power of electronic calculation far exceeds any mechanical calculation. But the

45

calculation is no longer intuitive, the steps are no longer displayed, and we no longer even use the decimal system we are familiar with, but are converted to binary.

Therefore, there are new languages called computer languages, which are used to input data and instructions to the computer and communicate with it. The instruction set that controls a computer is also called software, and writing software is called programming. Today programming is a required course for every engineer. A handheld calculator is also a must-have tool for every student.

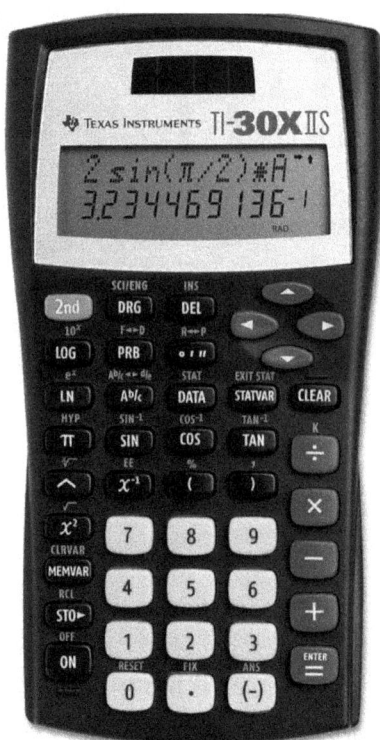

Figure 23. Texas Instruments scientific calculator 图 *23. 德州仪器科学计算机*

6   数学分支 Branches of Math

除算术以外, 数学还有很多分支, 如代数, 几何, 集合论, 微积分, 统计等等。 代数, 几何, 集合论中的一些基本概念, 小学里已经涉及。 这一章里我们将简单介绍初级数学中会用到的一些词汇。

In addition to arithmetic, there are many other branches of mathematics, such as algebra, geometry, set theory, calculus, statistics, and so on. Some of the basic concepts in algebra, geometry, and set theory are already covered in elementary school. In this chapter, we will briefly introduce some of the vocabulary used in basic math.

代数 Algebra

代数使用符号替代数字进行运算。通常我们会使用罗马字母, 如 x y z 来替代数字。这些符号通常称为变量或者未知数, 因为它们代表的数值可以是变化的和不确定的。

Algebra uses symbols instead of numbers to perform operations. Usually we use Roman letters, such as x y z to replace numbers. These symbols are usually called variables or indeterminate because the values they represent can be changing or unknown.

变量 Variable

变量是用字符来代表不确定的数字。变量常用来求解未知数, 比如姐姐比弟弟大 3 岁, 姐姐 12 岁, 弟弟多大? 假设弟弟的年龄是 x, $x + 3 = 12$, $x = 9$。

变量也常用来表示一些数量之间的关系，如圆周长 c 和半径 r 的关系：$c = 2 \pi r$。很多数学和物理的规律都用变量来表达，简单如 $x * 1 = x$ 对任何数 x 都成立。

Variables are characters used to represent uncertain numbers. Variables are often used to solve for unknowns, for example, sister is 3 years older than brother, sister is 12, how old is the brother? Suppose the brother's age is x, $x + 3 = 12$, $x = 9$.

Variables are also commonly used to represent relationships between quantities, such as the relationship between the circumference of a circle, c, and its radius, r: $c = 2 \pi r$. Many laws of math and physics are expressed in terms of variables, as simple as $x * 1 = x$ holds true for any number x.

常量 Constant

与变量相对的是常量。并不是所有被字符替代的量都是变量，有时我们也会用字符表示不变的数字，比如圆周率 $\pi = 3.14159\cdots\cdots$

还有一种情况，我们会用字符表示初始条件下已经确定的量。比如在方程 $a x + b = 0$ 中 a 和 b 是已知量，虽然他们可以是任意数量，但在这个方程中终是已经确定，所以是常量。只有 x 是未知，是变量。

The opposite of variable is constant. Not all quantities replaced by characters are variables. Sometimes we also use characters to represent invariant numbers, such as $\pi = 3.14159...$

In another case, we use characters to represent quantities that have been determined under initial conditions. For example, in the equation $a x + b = 0$, a and b are known quantities. Although they can be any number, they have been predetermined in this equation, so they are constants. Only x is unknown, therefore a variable.

表达式 Expression

数学表达式是指数字、变量、常量、运算符的组合，当然必须是有意义的组合。下面我们给几个简单的表达式的例子：

A mathematical expression is a combination of numeric numbers, variables, constants, and operators, but of course it must be a meaningful combination. Below we give a few examples of simple expressions:

$$1 + 2 * 3$$
$$(x + 1)(x + 2)$$
$$2x^2 + 3x + 4$$
$$x / y$$

一般来说，如果赋予表达式中的变量一定的数值，表达式的数值是可以计算的。

In general, the value of an expression is computable if the variables in the expression are given certain values.

方程 Equation

方程是指含有变量的等式，即两个表达式相等。解方程是要找到变量的值，使得等式成立。使得等式成立的变量值叫做解或者根。

一元一次方程，又叫线性方程，只含有一个未知数变量，未知数的次数为 1。通常形式是 $ax + b = 0$（a, b 为常数，$a \neq 0$）。只有一个解：$x = -b/a$。

例如 $5 - 3 = 2$ 是等式，但不是方程，因为不含有变量。

$5(x - 1) + 3 = 4 * 2$ 是方程。解方程的方法是首先将方程一步步简化。

An equation is an equality containing variables, i.e., two expressions are equal. Solving an equation is about finding the values of the variables that make the equation true. The values of the variables that make the equation true are called solutions or roots.

A linear equation of one variable contains only one unknown variable, and the degree of the unknown variable is 1. The usual form is $ax + b = 0$ (a, b are constants, $a \neq 0$). There is only one solution: $x = -b/a$.

For example, $5 - 3 = 2$ is an equality, but it is not an equation because it contains no variables.

$5(x - 1) + 3 = 4 * 2$ is an equation. The way to solve an equation is to simplify it step by step:

$$5(x - 1) + 3 = 4 * 2$$
$$5x - 5 + 3 = 8$$
$$5x - 2 = 8$$
$$5x = 8 + 2 = 10$$
$$x = 10/5 = 2$$

函数 Function

　　函数至少有两个变量，两个变量之间有一一对应的关系。如 $y = 2x + 1$，对应每一个 x 数值，都有一个 y 值。在小学阶段，函数的概念只限于一一对应和寻找规律，下面的表格就是一个例子（表 10）。

A function has at least two variables, and a one-to-one correspondence between the two variables. For example, $y = 2x + 1$, for each x-value, there is a y-value. At the elementary school level, the concept of a function is limited to one-to-one correspondence and finding patterns, as exemplified in the table below (Table 10).

Table 10. Find pattern **表 10.** 寻找规律

x	1	2	4	6
y	3	5	9	?

几何 Geometry

几何是研究空间的结构和性质的数学分支，比如点与点之间的距离，角度的关系，物体的形状等等。在这里我们只简单介绍最基本的几何词汇，而不涉及更深的计算、逻辑和证明。

Geometry is the branch of mathematics that studies the structure and properties of space, such as the distance between points, the relationship of angles, the shape of objects, and so on. Here we will briefly introduce only the most basic geometric vocabulary without going into deeper calculation, logic, and proof.

点线面 Point Line Plane

点是空间中的一个位置，没有长宽高。线是点的集合，常指向两个方向无限延伸的直线。面是一个无限延伸的二维的表面（图24）。

两点可以决定一个直线，三个不在同一条直线上的点决定一个平面。平面里的两条直线可以是相交的，也可以是永远不相交，后者称为平行线。

从一点开始只向一个方向无限延伸的线叫半直线，或叫射线。连接两点的直线叫线段。

A point is a position in space and has no length, width or height. A line is a collection of points, often referring to straight lines that extend infinitely in two directions. A plane is an infinitely extending two-dimensional surface (Figure 24).

Two points can determine a straight line, and three points that are not on the same straight line can determine a plane. Two straight lines in a plane may intersect or never intersect, the latter being called parallel lines.

A line that starts from a point and extends infinitely in only one direction is called a semi-straight line, or a ray. A straight line connecting two points is called a line segment.

Point 点   Line 线   Plane 平面

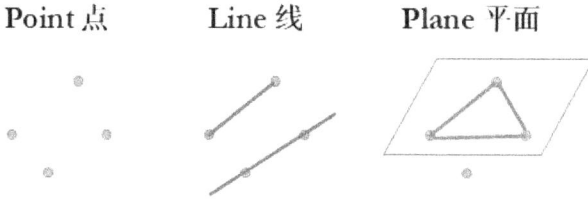

Figure 24. Point line and plane 图 24. 点线面

角 Angle

　　两个有共同起点的射线或者线段组成一个角，角的起点叫做顶点。角的大小称为角度。角也可以看成一个射线绕顶点的旋转，旋转一周回到起点位置是 360 度（360°）（图 25）。

Two rays or line segments with a common starting point form an angle, and the starting point of the angle is called a vertex. The size of an angle is called an angular measure or simply an angle. The angle can also be regarded as the rotation of a ray around the vertex. One complete rotation returning to the starting position is 360 degree (360°) (Figure 25).

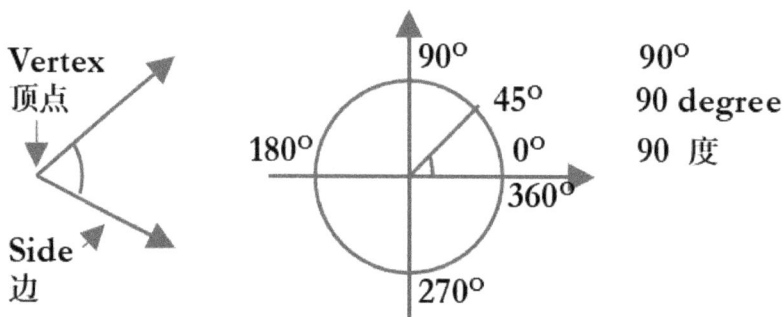

Figure 25. *Angle and degree* **图 25.** *角和角度*

角依据它们角度的大小和相对的位置，有各种各样的名称。图 26 列出了其中常见的一些名称。

Angles have various names depending on their angle size and relative position. Figure 26 lists some of the common names.

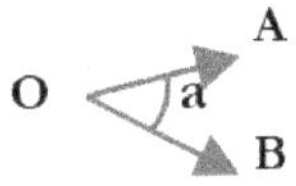 O ◁ a A B	Angle 角 $\angle AOB = \angle a$	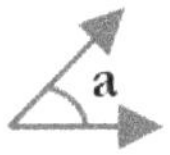	Acute Angle 锐角 ($\angle a < 90°$)
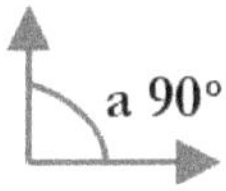 a 90°	Right Angle 直角 ($\angle a = 90°$)	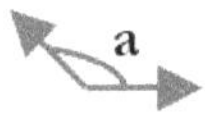	Obtuse Angle 钝角 ($\angle a > 90°$)
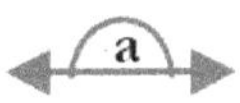 a	Straight Angle 平角 ($\angle a = 180°$)	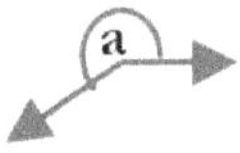 a	Reflex Angle 优角 ($\angle a > 180°$)
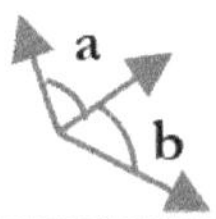 a b	Adjacent Angle 邻角	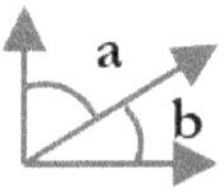 a b	Complementary Angle 余角 ($\angle a + \angle b = 90°$)
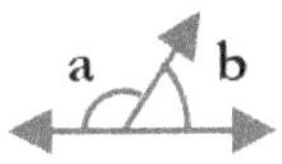 a b	Supplementary Angle 补角 ($\angle a + \angle b = 180°$)	 a b	Vertical Angle 对顶角 ($\angle a = \angle b$)

Figure 26. *Classification of angles* **图 26.** 角的分类

三角形 Triangle

三角形顾名思义有三个角，常见的分类见下图（图 27）。三角形也有三个边。

As the name suggests, a triangle has three angles. The common classification is shown in the figure below (Figure 27). A triangle also has three sides.

Types of Triangles 三角形的种类

Equilateral 等边

3 equal sides
3 等边
3 equal angles
3 等角 (60°)

Isosceles 等腰

2 equal sides
2 等边
2 equal angles
2 等角

Right 直角

Hypotenuse 斜边
1 right angle (90°)
1 直角 (90°)
90°

Scalene 不等边

No equal sides
无等边
No equal angles
无等角

Figure 27. Types of triangles **图 27.** 三角形的种类

多边形 Polygon

多边形是由多条边组成的封闭形状，三角形是其中最简单的一种。下图（图 28）里是一些常见的多边形。

A polygon is a closed shape made up of multiple sides, of which the triangle is the simplest. The picture below (Figure 28) shows some common polygons.

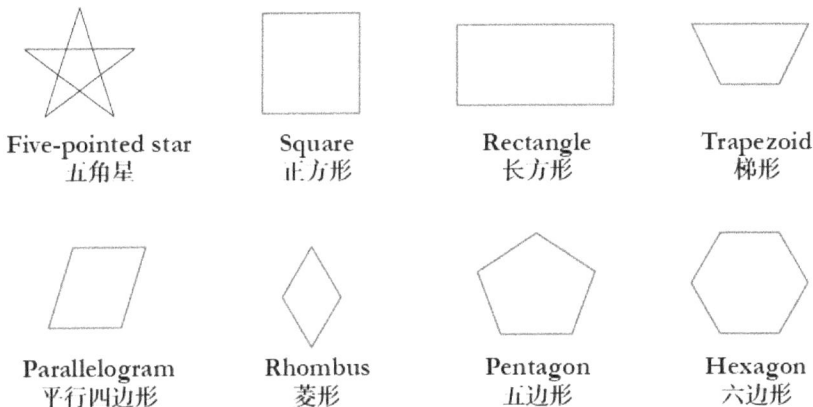

Figure 28. *Types of polygons* **图 28.** 多边形的种类

Five-pointed star 五角星	Square 正方形	Rectangle 长方形	Trapezoid 梯形
Parallelogram 平行四边形	Rhombus 菱形	Pentagon 五边形	Hexagon 六边形

曲线 Curve

任何连续的线条都是曲线，包括弧线、圆、椭圆（图 29）。直线也是曲线的一种。

Any continuous line is a curve, including arcs, circles, and ellipses (Figure 29). A straight line is also a type of curve.

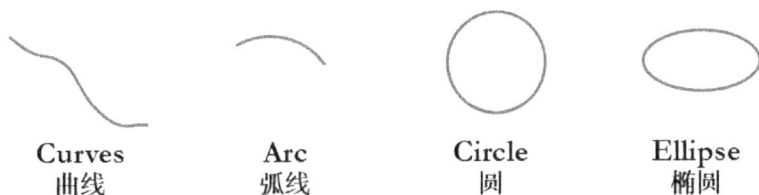

Curves 曲线	**Arc** 弧线	**Circle** 圆	**Ellipse** 椭圆

Figure 29. *Curves* **图 29.** 曲线

57

立体几何 Solid Geometry

如果一个物体的点不属于同一个平面，那么这个物体就是立体的，或者叫三维的。立体几何研究立体的分类和性质（图 30）。

If the points of an object do not belong to the same plane, then the object is three-dimensional, or solid shapes. Solid geometry studies the classification and properties of solids (Figure 30).

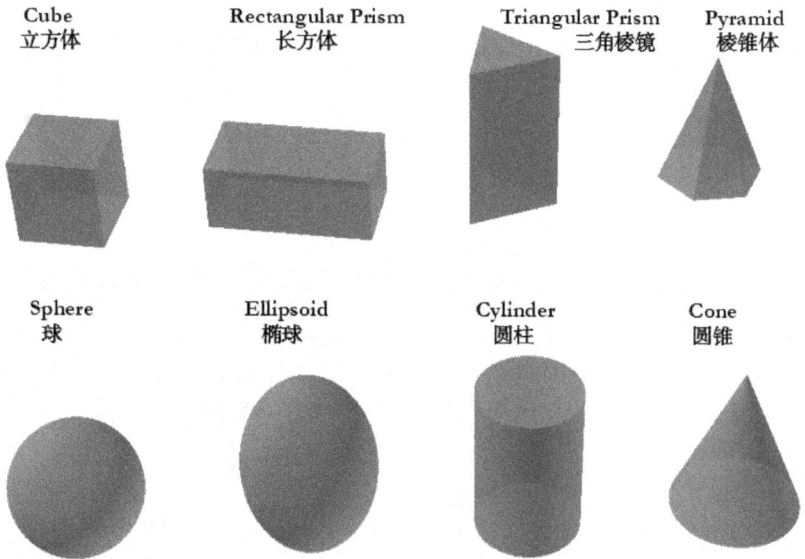

Figure 30. *Examples of solids* **图 30.** 立体的例子

周长面积体积 Perimeter Area Volume

描述一个物体，我们常常从其大小开始，简单如长宽高（图31）。对一个面其周长和面积，也常用来表示物体的大小；对三维物体，还需要体积。

To describe an object, we often start with its size, as simple as its length, width and height (Figure 31). For a surface, its perimeter and area are also used to describe the size of the object; for three-dimensional objects, the volume is also needed.

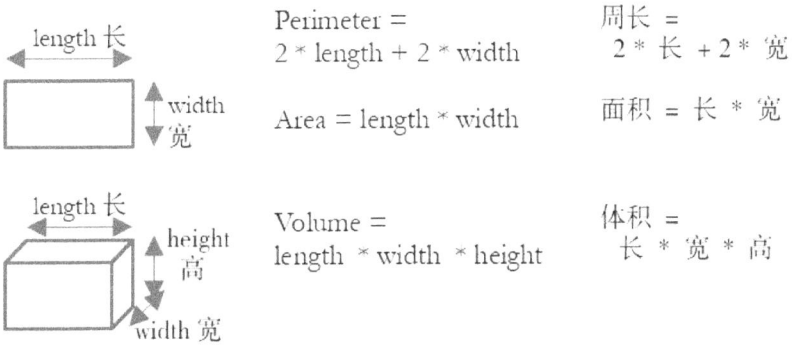

Perimeter =
2 * length + 2 * width

Area = length * width

Volume =
length * width * height

周长 =
2 * 长 + 2 * 宽

面积 = 长 * 宽

体积 =
长 * 宽 * 高

Figure 31. Perimeter area volume 图 31. 周长面积体积

圆的周长叫圆周长（图32），与圆的直径成正比关系。圆周长与直径的比是一个重要常数，圆周率，π。

The perimeter of a circle is called the circumference (Figure 32), which is directly proportional to the diameter of the circle. The ratio of a circle's circumference to its diameter is an important constant, PI, π.

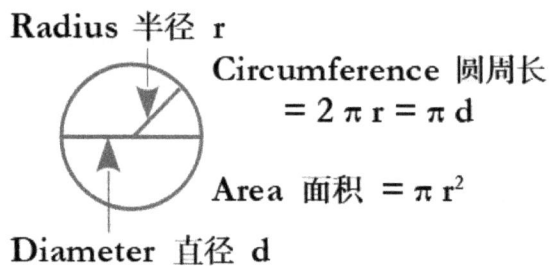

Radius 半径 r

Circumference 圆周长
$= 2\pi r = \pi d$

Area 面积 $= \pi r^2$

Diameter 直径 d

Figure 32. *Circumference and area* **图 32.** 圆周长和面积

集合论 Set Theory

我们常常在没有意识到的情况下谈论集合，例如整数就是一个数的集合。这里我们简单地引介集合的概念。

Often we talk about set without realizing it, for example integer is a set of numbers. Here is a brief introduction to the concept of set.

集合 Set

集合是一组物件形成的整体，集合中的物件称为元素或成员。元素属于集合，集合包含元素。常用 { } 表示集合，元素置于括号之中。下面是一系列集合的例子。

（1）A = {偶数}

（2）B = {0，2，4，6，8……}

（3）C = {红，黄，蓝}

（4）D = {2，4，5，2}

集合的元素可以是任何事物，可以是描述形的，也可以是列举形的。如上面的例子（1）集合 A 是描述形，（2）集合 B 是列举形，但两者是相同的集合，A = B。

集合中的元素的顺序无关紧要，重复的元素也只算成一个元素。如 {2，4，5，2} = { 2，4，5} = {5，4，2}。

A set is a group of objects that form a collection, and the objects in a set are called elements or members. Elements belong to the set (in the set), set contains the elements. Sets are often represented by { }, with the elements enclosed in parentheses. The following are examples of a series of sets.

(1) A = {even numbers}

(2) B = {0, 2, 4, 6, 8......}

(3) C = {red, yellow, blue}

(4) D = {2, 4, 5, 2}

The elements of a set can be anything, defined either by description or by listing. As in the above example (1) set A is defined by description and (2) set B is listing, but both are the same set, A = B.

The order of the elements in the set is irrelevant, and duplicates count as only one element. For example {2, 4, 5, 2} = { 2, 4, 5} = {5, 4, 2}.

子集 Subset

两个集合 A 和 B，如果集合 A 的所有元素都是集合 B 的元素，那么 A 是 B 的子集合，B 则是 A 的超集合（又称父集，母集）。写作 A ⊂ B，读作 A 包含于 B，或者 B 包含 A。

集合 A 是 A 自己的子集。

空集是指不含任何元素的集合。空集是任何集合 A 的子集。

例如：

偶数集合是整数集合的子集。

A = {3, 8, 9}, B = {2, 3, 5, 6, 8, 9}, A ⊂ B。

C = {苹果，梨子，香蕉}, D = {核桃，杏仁，栗子}, C 不是 D 的子集合。

Giving two sets A and B, if all the elements of set A are also elements of set B, then A is a subset of B, and B is a superset of A. Writes as A ⊂ B and reads as A is included in B, or B contains A.

The set A is a subset of A itself.

The empty set is a set that contains no elements. The empty set is a subset of any set A.

Example:

The set of even numbers is a subset of the set of integers.

A = {3, 8, 9}, B = {2, 3, 5, 6, 8, 9}, A ⊂ B.

C = {apples, pears, bananas}, D = {walnuts, almonds, chestnuts}, C is not a subset of D.

交集和并集 Intersection and Union

交集是两个集合 A B 共有的元素组成的集合，写作 A ∩ B。

并集是两个集合 A B 所有的元素组成的集合，写作 A ∪ B。

例如：

A = {1, 3, 8, 9}, B = {2, 3, 5, 6, 8, 9}，A ∩ B = {3, 8, 9}。

C = {苹果，梨子，香蕉}，D = {核桃，杏仁，栗子}，C∪D = { 苹果，梨子，香蕉，核桃，杏仁，栗子}。

The intersection is the set of elements common to two sets A B, written as A ∩ B.

The union is the set consisting of all the elements of two sets A B, written as A ∪ B.

Example:

A = {1, 3, 8, 9}, B = {2, 3, 5, 6, 8, 9}, A ∩ B = {3, 8, 9}.

C = {apple, pear, banana}, D = {walnut, almond, chestnut}, C ∪ D = {apple, pear, banana, walnut, almond, chestnut}.

文氏图 Venn Diagram

　　文氏图（韦恩图，Venn 图，图 33）常被用来表示集合之间的相互关系。通常用一个椭圆（或者圆）来代表一个集合，两个椭圆的共同部分表示两个集合的交集，两个椭圆的全部表示两个集合的并集。在下面的左图中（图 33，左图），集合 A 的颜色是深灰色，集合 B 的颜色是浅灰色，中间重叠的部分表示 A 和 B 的交集。在右图（图 33，右图），A 和 B 都被填成深灰色，同一种颜色代表 A 和 B 的合集。

Venn diagrams (Figure 33) are often used to represent interrelationships between sets. An ellipse (or circle) is usually used to represent a set, with the common part of the two ellipses representing the intersection of the two sets, and all parts representing the union of the two sets. In the left diagram below (Figure 33, left), set A is colored dark gray, set B is colored light gray, and the overlap in the middle represents the intersection of A and B. In the figure on the right (Figure 33, right), both A and B are filled with dark gray, and the same color represents the union of A and B.

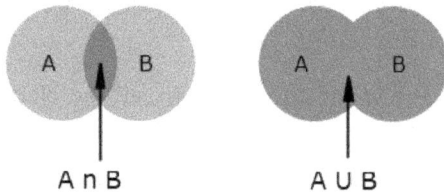

Figure 33. *Venn diagram* **图 33.** 文氏图

统计 Statistics

统计是收集和分析数据的学科。在这里我们只简单介绍数据集、平均值、中位数、众数、最大、和最小等概念。

数据集 一组数据的收集，如 {2，3，3，7，8，6，11，9}。

平均值 所有数据的和除以数据的个数。在上面的例子里，

平均值 = (2 + 3 + 3 + 7 + 8 + 6 + 11 + 9) / 8 = 49 / 8 = 6.125。

中位数 将所有数据按从小到大排列，取最中间的数。如果数据的个数是偶数，则取中间两个数的平均值。我们的例子从小到大排列是 {2，3，3，6，7，8，9，11}，数据的个数是偶数，中位数 = （6 + 7) / 2 = 6.5。

众数 众数是在数据集中出现次数最多的数，我们的例子中 3 出现了两次，其他都是一次，众数 = 3。数据集可以有不止一个众数。

最大值 数据集中最大的数。

最小值 数据集中最小的数。

Statistics is the discipline of collecting and analyzing data. Here we will only briefly introduce data sets, mean, median, mode, maximum, and minimum.

Dataset A collection of data such as {2, 3, 3, 7, 8, 6, 11, 9}.

Mean The sum of all the data divided by the number of data. In the example above,

mean = (2 + 3 + 3 + 7 + 8 + 6 + 11 + 9) / 8 = 49 / 8 = 6.125.

Median Arranges all the data from smallest to largest and takes the middle number. If the number of data is even, take the average of the two middle numbers. In our example, from smallest to largest, it is {2, 3, 3, 6, 7, 8, 9, 11}, the number of data is even, median = (6 + 7) / 2 = 6.5.

Mode Mode is the number that occurs the most times in the data set, in our example 3 occurs twice and everything else once, Mode = 3. A data set can have more than one mode.

Maximum The largest number in the data set.

Minimum The smallest number in the data set.

7　词汇表 Glossary

数学符号 Math Symbols

Symbol	English	中文	Symbol	English	中文
+	plus, add	加	=	equal	等于
−	subtract, minus	减	≠	not equal	不等于
* or × or ·	multiply, time	乘	≈	approximately	约等于
/ or ÷	divide	除	<	smaller than	小于
** or ^	exponent	乘方	>	greater than	大于
% or mod	modulo	取模	≤ or <=	not greater than	不大于，小于等于
%	percent	百分比	≥ or >=	not smaller than	不小于，大于等于
π	Pi	派	∈	is a member of	属于
( )	Parentheses	括号，小括号	∩	intersection	交集
[ ]	Square brackets	中括号，方括号	∪	union	并集
{ }	Curly brackets	大括号，花括号			

词汇 Vocabulary

字母顺序 Alphabetical Order

A

1D, one dimension	一维
2D, two dimension	二维
3D, three dimension	三维
Abacus	算盘
Absolute value	绝对值
Accuracy	精度
Acute angle	锐角
Acute triangle	锐角三角形
Add	加
Addend	加数
Addition	加法
Additive inverse	加法逆元; 相反数
Adjacent	相邻
Algebra	代数
Algorithm	算法
Ambiguity	歧义
Amount	量
Analog	模拟
Analyze	分析
Angle	角
Annual	每年
Answer	答案；回答
Ante meridian (a.m.)	上午
Apex	顶点
Application	应用；程序
Apply	运用
Approach	接近；方法
Arabic	阿拉伯
Arc	圆弧
Area	面积
Argument	自变量；论据
Arithmetic	算术
Array	数组
Ascending	上升；渐升
Associative property	结合特性;结合律
Assume	假如
Attribute	属性；归因于
Average	平均值
Axis (axes)	轴

B

Bar	杠
Bar graph	条形图
Base	底
Bi-	两个
Billion	十亿
Binary	二进制
Bisect	二等分
Border	边界
Borrow	借；退位
Bottom	底部
Brahmi	婆罗米
Branch	分支

C

Calculate	计算
Calculator	计算器
Calculus	微积分
Calendar	日历
Capacity	容量
Capital letter	大写字母
Carry	进位；携带
Cent	分

Centimeter (cm)	公分；厘米	Composite number	合数；合成数
Chart	图表	Computation	计算
Check	检查	Computer	计算机
Choice	选择	Concentric	同心
Choose	选取	Conclusion	结论
Circle	圆	Cone	圆锥
Circle graph	圆饼图	Congruent	全等
Circular	圆形	Congruent triangles	全等三角形
Circumference	圆周长	Conjecture	猜想
Classification	分类	Connect	连接
Clockwise	顺时针	Consecutive	相邻的；连续的
Closed	封闭	Constant	常数
Collection	收集	Construct	构造
Combine	结合	Contrast	对比
Common	共同	Convert	转换
Common denominator	公分母	Coordinate	坐标
Common factor	公约数；公因子	Corner	角落
Common multiple	公倍数	Correct	正确
Commutative property	交换律	Corresponding	对应的
Compare	比较	Corresponding angles	同位角
Comparison	比较	Count	数数
Compass	圆规	Counter	相反的
Complementary angles	余角	Counter Clockwise	反时针
		Cube	正方体
Composite number	合数；合成数	Cubic	立方；三次方
Computation	计算	Currency	货币
Computer	计算机	Curve	曲线
Concentric	同心	Customary	习惯
Conclusion	结论	Cylinder	圆柱
Cone	圆锥		
Congruent	全等		
Congruent triangles	全等三角形	**D**	
Conjecture	猜想	Data	数据
Connect	连接	Database	数据库
Consecutive	相邻的；连续的	Dataset	数据集
Constant	常数	Day	日；天
Construct	构造	Decagon	十边形
Compass	圆规	Decimal	十进制；小数
Complementary angles	余角	Decimal point	小数点

Decimeter	分米；十分之一米	Elapsed time	过去的
Decompose	分解	Element	元；元素
Decrease	减少	Elementary	基本的；初级的
Decreasing	减少；递减	Elementary School	小学
Definition	定义	Eleven	十一
Degree	度	Elimination	消元；消除
Denominator	分母	Ellipse	椭圆
Depth	深度	Ellipsoid	椭球
Descending	递减	Endpoint	终点
Diagram	图表	Equal (=)	等于
Diameter	直径	Equation	方程
Difference	差；差别	Equidistant	等距
Digit	数字；数位	Equilateral	等边多边形
Digital	数码的；电子的	Equivalent	等效；等价
Dime	一毛钱	Estimate	估量；估计值
Dimension	维	Estimation	估计
Distance	距离	Evaluate	计值；评估
Distributive property	分配律	Even number	偶数
Divide	除	Exam	考试
Dividend	被除数	Examine	检查
Divisibility	可除性	Example	例子
Divisible	可除的	Exchange	互换
Division	除法	Expand	展开
Divisor	除数；约数；因子	Experiment	实验
		Explain	解释
Dodecahedron	十二面体	Exploration	探索
Dollar ($)	元；美元	Explore	探求
Dot	点	Exponent	乘方
Double	加倍；两倍	Exponential	指数
Double check	再检查	Expression	表达式
		Extend	扩展
		Exterior	外部
		Exterior angles	外角
		Extra	多余；另加

E

e.g.	例如
Edge	棱；边
Eight	八
Eighteen	十八
Eighth	第八；八分之一
Eighty	八十

F

Face	面
Fact	事实
Factor	因子
Factorial	阶乘
Factoring	因子分解

False	错
Feet	英尺
Fewer than	比……少
Fifteen	十五
Fifth	第五；五分之一
Fifty	五十
Figure	图
Find	找到
Finger	手指
Finite	有限
First	第一
Five	五
Flip	翻转
Foot (ft)	英尺
For example	例如
Formula	公式
Formulate	规划；表达
Forty	四十
Four	四
Fourteen	十四
Fourth	第四；四分之一
Fraction	分数
Fraction bar	分数线
Frequency	频率
Function	函数
Fundamental theorem of arithmetic	算术基本定理

G

Gallon (gal)	加仑
Generate	生成
Geometric	几何的
Geometry	几何
Gram (g)	克
Graph	图；画图
Graph paper	坐标纸
Graphical	用图表示的
Greater	更大
Greatest	最大的
Greatest common	最大公约数
divisor (GCD)	
Greatest common factor (GCF)	最大公因子
Grid	方格
Group	组；分组
Guess	猜

H

Half	半
Halve	二分之一
Halving	对半分
Height	高
Heptagon	七边形
Hexagon	六边形
Higher	更高
Hindu	印度
Horizontal	水平的
Hour	小时
Hour hand	时针
Hundred	百
Hundredth	第一百；百分之一
Hypotenuse	斜边
Hypothesis	假说

I

i.e.	亦即；就是
Identical	相同的
Identify	找出
Identity	恒等式
Improper	不正当的
Improper fraction	假分数
Inch (in)	英寸
Incorrect	不正确
Increase	增加
Increasing	从小到大顺序
India	印度
Indeterminate	未知数

Inequality	不等式	Like Denominators	公分母
Infinite	无限	Likely	可能的
Inside	里面	Line	直线
Instance	例子	Line graph	线型图
Integer	整数	Line plot	折线图
Interior	内部	List	列出；清单
Input	输入	Liter (L)	公升；升
Interior angles	内角	Logic	逻辑
Interpret	解释	Long division	长除法
Intersect	相交		
Intersecting lines	相交线		
Intersection set	交集		
Invalid	无效	**M**	
Inverse	反向	Manipulation	操作；处理
Inverse operation	逆运算	Map	地图；标出；对应
Inverse property	逆向特性		
Invert	倒转	Marks	标记
Investigate	调查	Match	配对
Irrational number	无理数	Mathematics	数学
Irregular	不规则	Max	最大
Irrelevant	不相关	Maximum	最大
Isosceles triangle	等腰三角形	Mean	平均值
		Measure	测量；量度
		Measurement	测量
		Median	中位数
J K L		Meter (m)	公尺；米
Justify	证明；解释	Method	方法
Key	关键；重点	Metric system	公制
Key to a graph	图例	Metric units	公制单位
Kilogram (kg)	公斤	Micro	微
Kilometer (Km)	公里	Middle	中间
Label	标记	Midpoint	中点
Last	最后	Mile	英里
Least	最小	Milliliter (ml)	毫升
Least common denominator (LCD)	最小公分母	Millimeter (mm)	毫米
		Million	百万
		Min	最小
Least common multiple (LCM)	最小公倍数	Minimum	最小
		Minuend	被减数
Length	长度	Minus	减；负
Less	更少	Minute	分钟
Less than (<)	少于	Mixed number	带分数
Like	相似	Mode	众数

Modulo	取模	Obtuse triangle	钝角三角形
Month	月	Octagon	八边形
More	更	Odd number	奇数
More than (>)	大于(>)	One	一
Multiple	倍数；多个	Open	开放
Multiple Choice	多种选择	Operation	运算
Multiplicand	被乘数	Order	次序
Multiplication	乘法	Order of	运算次序
Multiplicative inverse	倒数	operations	
		Ordinal number	序数
Multiplier	乘数	Organize	组织
Multiply	乘	Organized	有序
		Origin	原点
		Ounce (oz)	盎司
		Output	输出；结果
		Outside	外面

N

Negative	负；不
Negative number	负数
Next	下一个
Nickel	五分钱
Nine	九
Nineteen	十九
Ninety	九十
Ninth	第九；九分之
Nonagon	九边形
Nonstandard	非标准
Not	不
Null	零；无
Number	数字
Number theory	数论
Numeral	数位；数字系统
Numeration	计数方法
Numerator	分子
Numeric	数；数字的
Numeric expression	数字表达式

P

Parabola	抛物线
Parallel	平行
Parallelogram	平行四边形
Parity	奇偶性
Part	部分
Pattern	模式
Penny	一分钱
Pentagon	五边形
Per	每个
Percent	百分比
Perimeter	周长
Perpendicular	垂直
Pi	圆周率；派
Pictograph	象形图；统计图表
Pint (pt)	品脱（pt）
Place value	位值
Plane	平面
Plot	绘图
Plus	加
Point	点
Polygon	多边形
Polyhedron	多面体

O

Object	物体
Obtuse angle	钝角

Positive	正；正确	Quarter	四分之一；25美分
Positive number	正数		
Possible	可能	Quarter hour	一刻钟
Post meridian (p.m.)	下午	Quarter year	一季度
		Question	问；问题
Pound (lb)	磅（lb）	Quiz	测试
Power	幂；乘方	Quotient	商
Predict	预测		
Prediction	预测		
Prefixes	前缀		
Previous	前一个		
Primary school	小学	**R**	
Prime factorization	质因子分解	Radius	半径
Prime number	质数	Range	范围；值域
Prism	棱镜	Rate	速率；率
Probability	概率；几率；可能性	Ratio	比；比率
		Rational number	有理数
Problem	问题	Rationale	依据
Process	过程	Ray	射线
Product	积；乘积	Real number	实数
Program	程序	Reasonable	合理
Programming	编程	Reciprocal	倒数
Proof	证明	Rectangle	长方形；矩形
Proper	正当的	Rectangular Prism	长方体
Proper fraction	真分数	Recurring	反复出现；循环
Property	特性	Reduce	简化；减少
Proportional	正比；成正比	Reduction	减少；约分
Protractor	量角器；分度规	Reference	参考；参照
Prove	证明	Reference frame	参考系；坐标系
Pyramid	棱锥体	Reflect	反射
Pythagorean Theorem	勾股定理	Reflex angle	优角
		Regroup	重组
		Regular	正；正常的
		Relation	关系
		Relevant	相关
		Remainder	余数
Q		Repeat	重复
Quadrangle	四边形	Repeating decimal	循环小数
Quadrant	象限；四分之一圆	Representation	表示
		Review	回顾;复习
Quadrilateral	四边形	Rhombus	菱形；斜方块
Quantity	量	Right	正确；右
Quart	夸脱	Right angle	直角

Roman	罗马	Solid	实
Rotation	旋转	Solution	解
Round down	下舍入；四舍	Solve	解；解答
Round up	上舍入；五入	Sort	分类
Rounding	四舍五入	Space	空间
Rule	规则	Special case	特例
Ruler	尺	Specific	特定
		Speed	速度
		Sphere	球体
S		Square	平方；正方形
Sample	样本	Standard	标准
Scale	比例；规模	Statement	陈述
Scalene triangle	不等边三角形	Statistics	统计
Scientific notation	科学记数法	Step	步骤；阶梯
Seasons	季节	Straight	直
Second	秒	Straight angle	平角
Sector	扇形	Subscript	下角文字；下标
Segment	段	Substitute	替换；代入
Semester	学期	Subtotal	部分和
Semicircle	半圆	Subtract	减
Semi-straight line	半直线	Subtraction	减法
Separately	分别	Subtrahend	减数
Separator	分隔号	Sum	和
Set	集合	Summarize	总结
Seven	七	Summary	概述
Seventeen	十七	Superscript	上角文字；上标
Seventh	第七；七分之一	Supplementary	补角
Seventy	七十	angles	
Shape	形状	Suppose	假如
Share	份	Survey	调查；测量
Show	说明	Swap	互换
Side	边	Symbols	符号
Significant figures	有效数字	Symmetry	对称
Similar	相似		
Simplify	简化		
Simultaneously	同时	**T**	
Six	六	Table	表
Sixteen	十六	Tall	高
Sixty	六十	Tally	计数
Slope	斜坡；斜率	Technique	技术；技巧
Small	小	Ten	十
Software	软件		

Tenth	第十；十分之一	Variable	变量
Term	项	Venn diagram	维恩图；温氏图；文氏图
Terminate	终结		
Theorem	定理	Verify	验证
Third	第三，三分之一	Vertex	顶点
Thirteen	十三	Vertical	垂直的
Thirty	三十	Vertical angles	对顶角
Thousand	千	Vertices	顶点
Thousandth	第一千；千分之一	Volume	容量；体积
		Week	星期
Three	三	Whole	整体
Three-dimensional	三维；立体	Whole number	整数
Time	时间；乘	Width	宽；宽度
Ton	吨	Yard	码
Top	上方；顶部	Zero	零；零点
Total	总和		
Translate	平移		
Trapezoid	梯形		
Tri-	三个		
Trial	实验		
Trial and error	试错		
Triangle	三角形		
Triangular Prism	三角棱角		
Trillion	兆；万亿		
Triple	三倍		
True	是		
Turn	转；圈		
Twelve	十二		
Twenty	二十		
Two	二		
Type	种类		

U-Z

Union set	并集
Unique	唯一
Unit	单位
Unlike denominators	异分母
Valid	有效；正确
Value	数值

笔画排序 Order by Number of Strokes

1

一	One
一分钱	Penny
一刻钟	Quarter hour
一季度	Quarter year; season
一毛钱	Dime
一维	1D, one dimension

十二面体	Dodecahedron
十五	Fifteen
十亿	Billion
十八	Eighteen
十六	Sixteen
十分之一	Tenth
十四	Fourteen
十边形	Decagon
十进制	Decimal

2

七	Seven
七分之一	Seventh
七十	Seventy
七边形	Heptagon
九	Nine
九分之一	Ninth
九十	Ninety
九边形	Nonagon
二	Two
二分之一	Halve
二十	Twenty
二等分	Bisect, half
二维	2D, two dimension
二进制	Binary
八	Eight
八分之一	Eighth
八十	Eighty
八边形	Octagon
几何	Geometry
几何的	Geometric
几率	Probability
十	Ten
十一	Eleven
十七	Seventeen
十三	Thirteen
十九	Nineteen
十二	Twelve

3

万亿	Trillion
三	Three
三个	Tri-
三倍	Triple
三分之一	Third
三十	Thirty
三维	3D, three dimension
三角形	Triangle
三角棱角	Triangular Prism
上升	Ascending
上午	Ante meridian (a.m.)
上方；顶部	Top
上标	Superscript
上舍入	Round up
上角文字	Superscript
下一个	Next
下午	Post meridian (p.m.)
下标	Subscript
下舍入	Round down
下角文字	Subscript
习惯	Customary
千	Thousand
千分之一	Thousandth
大于(>)	More than (>)
大写字母	Capital letter
小	Small

小学	Elementary School	六	Six
小学	Primary school	六分之一	Sixth
小数	Decimal	六十	Sixty
小数点	Decimal point	六边形	Hexagon
小时	Hour	内角	Interior angles
		内部	Interior
		分	Cent
4		分别	Separately
不	Not; No	分子	Numerator
不正当的	Improper	分度规	Protractor
不正确	Incorrect	分支	Branch
不相关	Irrelevant	分数	Fraction
不等式	Inequality	分数线	Fraction bar
不等边三角形	Scalene triangle	分析	Analyze
不规则	Irregular	分母	Denominator
中位数	Median	分米	Decimeter
中点	Midpoint	分类	Sort;
中间	Middle		Classification
互换	Exchange, swap	分组	Group
五	Five	分解	Decompose
五入	Round up	分配律	Distributive
五分之一	Fifth		property
五分钱	Nickel	分钟	Minute
五十	Fifty	分隔号	Separator
五边形	Pentagon	勾股定理	Pythagorean
从小到大顺序	Increasing		Theorem
元；元素	Element	升	Liter (L)
元；美元	Dollar ($)	反向	Inverse
公倍数	Common multiple	反复出现	Recurring
公分	Centimeter (cm)	反射	Reflect
公分母	Common	反时针	Counter Clockwise
	denominator	少于	Less than (<)
公分母	Like Denominators	尺	Ruler
公制	Metric system	开放	Open
公制单位	Metric units	手指	Finger
公升	Liter (L)	文氏图	Venn diagram
公尺	Meter (m)	方格	Grid
公式	Formula	方法	Method; Approach
公斤	Kilogram (kg)	方程	Equation
公约数；公因子	Common factor	无	Null; None
公里	Kilometer (Km)	无效	Invalid
		无理数	Irrational number

无限	Infinite	印度	India, Hindu
天	Day	可能	Possible
日	Day	可能的	Likely
日历	Calendar	可能性	Probability
月	Month	可除性	Divisibility
比；比率	Ratio	可除的	Divisible
比……少	Fewer than; Less than	未知数	Indeterminate
比例	Scale	右	Right
比较	Compare; Comparison	四	Four
		四分之一	Fourth; Quarter
水平的	Horizontal	四十	Forty
立体	Three-dimensional	四舍	Round down
计值	Evaluate	四舍五入	Rounding
计数	Tally	四边形	Quadrangle, Quadrilateral
计数方法	Numeration		
计算	Calculate, computation	处理	Manipulation
		外角	Exterior angles
计算器	Calculator	外部	Exterior
计算机	Computer	外面	Outside
边	Side；Edge	对半分	Halving
边界	Border	对应的	Corresponding
长度	Length	对比	Contrast
长方体	Rectangular Prism	对称	Symmetry
长方形	Rectangle	对应	Map; Corresponding
长除法	Long division		
		对顶角	Vertical angles
		平均值	Average; Mean
		平方	Square
		平移	Translate
5		平行	Parallel
代入	Substitute	平行四边形	Parallelogram
代数	Algebra	平角	Straight angle
加	Add; Plus	平面	Plane
加仑	Gallon (gal)	正；正常的	Regular, normal
加倍	Double	正；正确	Positive
加数	Addend	正当的	Proper
加法	Addition	正数	Positive number
加法逆元	Additive inverse	正方体	Cube
半	Half	正方形	Square
半圆	Semicircle	正比	Proportional
半径	Radius	正确	Correct; Right; Valid
半直线	Semi-straight line		
		生成	Generate

用图表示的	Graphical	曲线	Curve
立方	Cubic	有序	Organized
过去的	Elapsed time	有效	Valid
过程	Process	有效数字	Significant figures
		有理数	Rational number
		有限	Finite
		杠	Bar

6

交换律	Commutative property	次序	Order
		百	Hundred
		百万	Million
交集	Intersection set	百分之一	Hundredth; Percent
份	Share		
众数	Mode	百分比	Percent
优角	Reflex angle	米	Meter (m)
兆	Trillion	约分	Reduction
全等	Congruent	约数	Divisor
全等三角形	Congruent triangles	考试	Exam
		自变量	Argument
共同	Common	论据	Argument
关系	Relation	负	Negative; Minus
关键	Key	负数	Negative number
再检查	Double check	运用	Apply
列出	List	运算	Operation
合数；合成数	Composite number	运算次序	Order of operations
合理	Reasonable		
同位角	Corresponding angles	进位	Carry
		连接	Connect
同心	Concentric	连续的	Consecutive
同时	Simultaneously	问；问题	Question
回顾	Review	问题	Problem
因子	Factor ; divisor	阶乘	Factorial
因子分解	Factoring	阶梯	Step
地图	Map		
多个	Multiple		
多余	Extra		
多种选择	Multiple Choice	**7**	
多边形	Polygon	两个	Bi-
多面体	Polyhedron	两倍	Double
并集	Union set	估计	Estimation
异分母	Unlike denominators	估量；估计值	Estimate
		位值	Place value
扩展	Extend	体积	Volume
收集	Collection	余数	Remainder

余角	Complementary angles	底部	Bottom
克	Gram (g)	总和	Total
初级的	Elementary	总结	Summarize
吨	Ton	构造	Construct
坐标	Coordinate	歧义	Ambiguity
坐标系	Reference frame	物体	Object
坐标纸	Graph paper	画图	Graph
夸脱	Quart	直	Straight
序数	Ordinal number	直径	Diameter
应用	Application	直线	Line
形状	Shape	直角	Right angle
找出	Identify		
找到	Find		
技术；技巧	Technique		
折线图	Line plot	**8**	
抛物线	Parabola	事实	Fact
时针	Hour hand	例子	Example; Instance
时间	Time		
更	More	例如	For example; e.g.
更大	Greater	依据	Rationale
更少	Less	函数	Function
更高	Higher	单位	Unit
条形图	Bar graph	参考；参照	Reference
步骤	Step	参考系	Reference frame
每个	Per	取模	Modulo
每年	Annual	变量	Variable
补角	Supplementary angles	周长	Circumference
		周长	Perimeter
角	Angle	和	Sum
角落	Corner	图	Figure; graph
证明	Justify; Proof; Prove	图例	Key to a graph
		图表	Chart
评估	Evaluate	垂直	Perpendicular
里面	Inside	垂直的	Vertical
阿拉伯	Arabic	奇数	Odd number
陈述	Statement	奇偶性	Parity
季节	Seasons	质因子分解	Prime factorization
学期	Semester	质数	Prime number
实	Solid	软件	Software
实数	Real number	退位	Borrow; regroup
实验	Experiment; Trial	逆向特性	Inverse property
底	Base	逆运算	Inverse operation

选取	Choose	恒等式	Identity
选择	Choice	指数	Exponential
非标准	Nonstandard	星期	Week
顶点	Apex; vertex; vertices	是	True
码	Yard	标准	Standard
空间	Space	标记	Label
线型图	Line graph	标记	Marks
组	Group	段	Segment
组织	Organize	派	Pi
终点	Endpoint	测试	Quiz
终结	Terminate	测量	Measure
罗马	Roman	测量	Measurement
英寸	Inch (in)	测量	Survey
英尺	Feet	点	Dot; point
英尺	Foot ft)	相交	Intersect
英里	Mile	相交线	Intersecting lines
范围	Range	相似	Similar; Like
表	Table	相关	Relevant
表示	Representation	相反数	Additive inverse
表达	Formulate; express	相反的	Counter
表达式	Expression	相同的	Identical
规划	Formulate	相邻	Adjacent
规则	Rule	相邻的	Consecutive
规模	Scale	真分数	Proper fraction
试错	Trial and error	科学记数法	Scientific notation
货币	Currency	秒	Second
定理	Theorem	结合	Combine
定义	Definition	结合特性; 结合律	Associative property
		结果	Result
		结论	Conclusion
		绘图	Plot
9		绝对值	Absolute value
前一个	Previous	统计	Statistics
前缀	Prefixes	说明	Show
厘米	Centimeter (cm)	转	Turn
品脱（pt）	Pint (pt)	转换	Convert
复习	Review	轴	Axis (axes)
封闭	Closed	速度	Speed
差；差别	Difference	速率	Rate
带分数	Mixed number	重复	Repeat
度	Degree	重点	Key
		重组	Regroup

钝角	Obtuse angle	消元；消除	Elimination
钝角三角形	Obtuse triangle	特例	Special case
除	Divide	特定	Specific
除数	Divisor	特性	Property
除法	Division	盎司	Ounce (oz)
面	Face	矩形	Rectangle
种类	Type	积	Product
面积	Area	被乘数	Multiplicand
项	Term	被减数	Minuend
顺时针	Clockwise	被除数	Dividend
		调查	Investigate; survey
		递减	Decreasing
		递减	Descending
10		逻辑	Logic
乘	Multiply; time	部分	Part
乘数	Multiplier	部分和	Subtotal
乘方	Power; Exponent	配对	Match
乘法	Multiplication	预测	Predict, prediction
乘积	Product	验证	Verify
倍数	Multiple	高	Height, tall
倒数	Multiplicative inverse	射线	Ray
倒数	Reciprocal		
倒转	Invert		
借	Borrow	**11**	
值域	Range	假分数	Improper fraction
原点	Origin	假如	Assume
圆	Circle	假如	Suppose
圆周率	Pi	假说	Hypothesis
圆周长	Circumference	偶数	Even number
圆弧	Arc	减	Subtract; Minus
圆形	Circular	减；负	Minus
圆柱	Cylinder	减少	Decrease
圆锥	Cone	减少	Reduction; reduce
圆规	Compass	减法	Subtraction
圆饼图	Circle graph	减数	Subtrahend
容量	Capacity	唯一	Unique
容量	Volume	商	Quotient
宽；宽度	Width	圈	Turn
展开	Expand	基本的	Elementary
扇形	Sector	婆罗米	Brahmi
替换	Substitute	常数	Constant
样本	Sample	排列	Arrange

探求	Explore	循环	Recurring
探索	Exploration	循环小数	Repeating decimal
接近	Approach	棱	Edge
斜坡；斜率	Slope	棱椎体	Pyramid
斜方块	Rhombus	棱镜	Prism
斜边	Hypotenuse	椭圆	Ellipse
旋转	Rotation	椭球	Ellipsoid
梯形	Trapezoid	温氏图	Venn diagram
检查	Check	程序	Program;
检查	Examine		Application
毫升	Milliliter (ml)	第一	First
毫米	Millimeter (mm)	第一千	Thousandth
深度	Depth	第一百	Hundredth
清单	List	第七	Seventh
渐升	Ascending	第三	Third
猜	Guess	第九	Ninth
猜想	Conjecture	第五	Fifth
率	Rate	第八	Eighth
球体	Sphere	第六	Sixth
符号	Symbols	第十	Tenth
维	Dimension	第四	Fourth
维恩图	Venn diagram	等于	Equal (=)
菱形	Rhombus	等效；等价	Equivalent
		等腰三角形	Isosceles triangle
		等距	Equidistant
		等边多边形	Equilateral
12		答案, 回答	Answer
最后	Last	编程	Programming
最大	Max; Maximum	象形图	Pictograph
最大公因子	Greatest common factor (GCF)	象限	Quadrant
最大公约数	Greatest common divisor (GCD)	距离	Distance
		量	Amount; Quantity
最大的	Greatest	量度	Measure
最小	Least; Min; Minimum	量角器	Protractor
		锐角	Acute angle
最小公倍数	Least common multiple (LCM)	锐角三角形	Acute triangle
最小公分母	Least common denominator (LCD)	集合	Set
属性；归因于	Attribute	**13**	
幂	Power	微	Micro

微积分	Calculus	增加	Increase
数；数字的	Numeric	磅（lb）	Pound (lb)
数位；数字系统	Numeral	操作	Manipulation
数值	Value	整体	Whole
数字	Number	整数	Integer
数字；数位	Digit	整数	Whole number
数字表达式	Numeric expression	翻转	Flip
数学	Mathematics		
数据	Data		
数据库	Database		
数据集	Dataset		
数数	Count		
数码的	Digital		
数组	Array		
数论	Number theory		
概率	Probability		
概述	Summary		
模式	Pattern		
简化	Simplify; reduce		
解	Solution		
解；解答	Solve		
解释	Interpret; Explain; Justify		
输入	Input		
输出	Output		
错	False		
零	Null		
零；零点	Zero		
频率	Frequency		

14+

模拟	Analog
算术	Arithmetic
算术基本定理	Fundamental theorem of arithmetic
算法	Algorithm; method
算盘	Abacus
精度	Accuracy

索引| Index

www.ingramcontent.com/pod-product-compliance
Lightning Source LLC
Chambersburg PA
CBHW071925020426
42331CB00010B/2726